Axiomatic Thinking I

VIII. Ordentliche Sitzung

der

Schweizerischen Mathematischen Gesellschaft

gemeinsam mit der

99. Jahresversammlung der Schweiz. Naturforschenden Gesellschaft

in Zürich

Dienstag, den 11. September 1917, im Hörsaal 304 der Universität

Vormittags punkt 8 Uhr

Vorträge und Mitteilungen der Herren:

1. *A. Emch* (Urbana U.S.A.): Ueber ebene Kurven, welche die n. Einheitswurzeln in der Ebene zu reellen Brennpunkten haben.
2. *G. Pólya* (Zürich): Arithmetische Eigenschaften der Reihenentwicklungen rationaler Funktionen.
3. *F. Gonseth* (Zürich): Un théorème sur deux ellipsoides confocaux.
4. *L. Kollros* (Zürich): Propriétés métriques des courbes algébriques.
5. *O. Spiess* (Basel): Ein Satz über rationale Funktionen.
6. *A. Hurwitz* (Zürich): Verallgemeinerung des Pohlkeschen Satzes.
7. *C. Carathéodory* (Göttingen): Ueber die geometrische Behandlung der Extrema von Doppelintegralen.

ERFRISCHUNGSPAUSE.

8. *D. Hilbert* (Göttingen): Axiomatisches Denken.
9. *A. Speiser* (Zürich): Ueber den Klassenkörper.
10. *S. Bays* (Fribourg): Une preuve directe que les systèmes triples de Kirkman et de Netto sont les seuls systèmes de triples de Steiner existants pour 13 éléments.
11. *L. G. Du Pasquier* (Neuchâtel): Sur un point de la théorie des nombres hypercomplexes.
12. *H. Berliner* (Bern): Ueber ein geometrisches Gesetz der infiniten Pluralität.
13. *K. Merz* (Chur): Quadratische Transformation einer Kollineation.
14. *G. Pólya* (Zürich): Ganzwertige Polynome in algebraischen Zahlkörpern.
15. *L. G. Du Pasquier* (Neuchâtel): Une nouvelle formule d'interpolation dans la théorie mathématique de la population.

NB. Die Herren Vortragenden werden ersucht, für die Dauer ihrer Vorträge nicht mehr als 20 Minuten in Aussicht zu nehmen und dem Sekretär einen Auszug ihrer Mitteilungen noch vor Schluss der Tagung abzugeben.

Gemeinsames Mittagessen um 1 Uhr im Hotel „Pelikan"

Im Anschluss daran sollen die Vereinsgeschäfte erledigt werden:

Abnahme der Jahresrechnung und Neuwahl des Vorstandes.

An die Mitglieder der Schweiz. Mathematischen Gesellschaft!

Sehr geehrte Herren Kollegen, wir unterbreiten Ihnen die reichhaltige Tagesordnung unserer ordentlichen Jahresversammlung und bitten um zahlreiche Teilnahme an dieser Tagung, welche eine besondere wissenschaftliche Bedeutung erhält durch den Vortrag des Herrn Prof. Hilbert aus Göttingen, der auf eine Einladung des Vorstandes über eine wissenschaftliche Methode sprechen wird, die ihm eine ausschlaggebende Förderung verdankt.

Der Vorstand der Schweiz. Mathematischen Gesellschaft:

Der Präsident: Prof. Dr. M. Grossmann (Zürich).

Der Vizepräsident: Prof. Dr. M. Plancherel (Fribourg).

Der Sekretär: Prof. Dr. L. Crelier (Biel-Bern).

Programme of the meeting of the Swiss Mathematical Society, September 11, 1917, in Zurich, Switzerland.
Courtesy of the Schweizerische Mathematische Gesellschaft.

Fernando Ferreira · Reinhard Kahle ·
Giovanni Sommaruga
Editors

Axiomatic Thinking I

Editors
Fernando Ferreira
Departamento de Matemática
Faculdade de Ciências
Universidade de Lisboa
Lisboa, Portugal

Reinhard Kahle
Carl Friedrich von Weizsäcker-Zentrum
Universität Tübingen
Tübingen, Germany

Giovanni Sommaruga
Department of Mathematics
ETH Zurich
Zurich, Switzerland

ISBN 978-3-030-77659-6 ISBN 978-3-030-77657-2 (eBook)
https://doi.org/10.1007/978-3-030-77657-2

Mathematics Subject Classification: 03-06, 03-03, 03A05, 03A10

This Springer imprint is published by the registered company Springer Nature Switzerland AG
The registered company address is: Gewerbestrasse 11, 6330 Cham, Switzerland

Preface

This book originates with two events commemorating the centenary of David Hilbert's seminal talk on *Axiomatic Thinking* (*Axiomatisches Denken*) which he delivered on September 11, 1917, at Zurich University for the Swiss Mathematical Society. This talk marks arguably the birth of proof theory as it was conceived by David Hilbert in the 1920s. It makes clear that the formalistic endeavor which one may find in the development of mathematical logic by the Hilbert school is, at best, a technical ingredient of a much larger enterprise which attempts to base every science deserving this predicate on a transparent *framework of concepts* (*Fachwerk von Begriffen*), developed and investigated by the axiomatic method.

On September 14–15, 2017, a joint meeting of the Swiss Mathematical Society and the Swiss Society for Logic and Philosophy of Science on *Axiomatic Thinking* took place at the University of Zurich, Switzerland, the place where Hilbert had spoken 100 years ago. It was followed, on October 11–14, 2017, by a conference on the same topic at the *Academia das Ciências de Lisboa* and *Faculdade de Ciências e Tecnologia da Universidade Nova de Lisboa* in Lisbon which was also the annual meeting of the *Académie Internationale de Philosophie des Sciences*. This meeting included a *Panel Discussion on the Foundations of Mathematics* with Peter Koellner, Michael Rathjen, and Mark van Atten as invited panelists.

The current volumes contain contributions of speakers of both meetings and also papers by other researchers in the field. In accordance with the broad range of topics addressed by Hilbert, the articles in Vol. I focus on reflections on the *History and Philosophy* of Axiomatic Thinking; Vol. II provides in Part I examples of developments of axiomatic thinking in *Logic*, especially in Proof Theory, inspired by Hilbert's ideas; Part II is concerned with applications of the axiomatic method in *Mathematics*; and Part III addresses the use of the axiomatic method in *other sciences*, namely Computer Science, Physics, and Theology.

Our dear friend Thomas Strahm, an inspired logician, followed the development of this book closely. But sadly he is not here to see its publication. Thomas Strahm died at the end of April, 2021. We dedicate this book to him—an excellent logician, and even more a kind, sensitive, humorous and wonderful friend.

Lisboa, Portugal
Tübingen, Germany
Zurich, Switzerland
March 2021

Fernando Ferreira
Reinhard Kahle
Giovanni Sommaruga

Acknowledgements

The editors are grateful to all speakers in Zurich and Lisbon who contributed to the success of the two meetings, as well as to the meetings' co-organizers, Thomas Kappeler and Viktor Schroeder in Zurich, and Gerhard Heinzmann, João Cordovil, João Enes, Mirko Engler, António Fernandes, Gilda Ferreira, Emanuele Frittaion, Nuno Jerónimo, Isabel Oitavem, Cecília Perdigão, and Gabriele Pulcini in Lisbon. The meeting in Zurich was supported by the Swiss Mathematical Society SMS, the Swiss Society for Logic and Philosophy of Science SSLPS, the Swiss Academy of Sciences SCNAT, and the Institute of Mathematics, University of Zurich, which is gratefully acknowledged. Equally, many thanks for the support of the conference in Lisbon to the Académie Internationale de Philosophie des Sciences AIPS; the Academia das Ciências de Lisboa; the Centro Internacional de Matemética, CIM; and the following research centers: Centro de Matemética, Aplicações Fundamentais e Investigação Operacional, CMAF-CIO; Centro de Matemática e Aplicações, CMA; Centre for Philosophy of Sciences of the University of Lisbon, CFCUL. The Portuguese Science Foundation (Fundação para a Ciência e a Tecnologia, FCT) supported the conference through the projects, UID/FIL/00678/2013, UID/MAT/00297/2013, UID/MAT/04561/2013, and PTDC/MHC-FIL/2583/2014 (*Hilbert's 24th Problem*). For the preparation of the current volumes, the first editor acknowledges the support by the Portuguese Science Foundation (UIDB/04561/2020) and the second editor was also supported by the Udo Keller Foundation.

Pro domo, we'd like to acknowledge Reinhard's enormous efforts and drive without which this book might still be a mere project. (Fernando and Giovanni).

Contents

Contents Overview of Vol. 2

Editors and Contributors

About the Editors

Fernando Ferreira is Professor of Mathematics at Universidade de Lisboa. He received his Ph.D. at Pennsylvania State University in 1988 under the direction of Stephen Simpson. He was a Fulbright Scholar at Harvard University (Spring 2004) and Tinker Visiting Professor at Stanford University (Fall 2009). He has written papers in weak systems of arithmetic and analysis, proof theory (especially functional interpretations) and philosophy and foundations of mathematics. He also wrote two papers on the problem of truth in Parmenides and Plato. He is a member of Academia das Ciências de Lisboa.

Reinhard Kahle is Carl Friedrich von Weizsäcker Professor for Philosophy and History of Science at the University of Tübingen. Before he holds professorships in Mathematics at the University of Coimbra and at the Universidade Nova in Lisbon, at the end as full professor for Mathematical Logic. He is a fellow of the *Académie Internationale de Philosophie des Sciences*. His main research interests are proof theory and the history and philosophy of modern mathematical logic, in particular, of the Hilbert School. He has (co-)edited more than ten books and special issues as, for instance, Gentzen's Centenary: The quest for consistency (Springer, 2015) and The Legacy of Kurt Schütte (Springer, 2020), both together with Michael Rathjen.

Giovanni Sommaruga did his studies in philosophy and philosophical and mathematical logic at the University of Freiburg (Switzerland), Stanford University, and the University of Siena. In 1996 he became assistant professor in logic and philosophy of science at the Albert Ludwig University Freiburg (Germany), and since 2008 he has been senior scientist in philosophy of the formal sciences (logic, mathematics, theoretical computer science) at ETH Zurich. His main research interests are in philosophy and foundations of mathematics, in the history of mathematical logic, and more recently in the history and philosophical issues concerning computability and information in theoretical computer science.

Contributors

Evandro Agazzi Center for Bioethics of the Panamerican University of Mexico City, Universities of Genoa, Genoa, Italy

Paul Bernays Eidgenössische Technische Hochschule, Zurich, Switzerland

Erwin Engeler Eidgenössische Technische Hochschule, Zurich, Switzerland

Fernando Ferreira Universidade de Lisboa, Lisboa, Portugal

José Ferreirós Universidad de Sevilla, Seville, Spain

David Hilbert Mathematisches Institut, Universität Göttingen, Göttingen, Germany

Reinhard Kahle Carl Friedrich von Weizsäcker-Zentrum, Universität Tübingen, Tübingen, Germany;
CMA, FCT, Universidade Nova de Lisboa, Caparica, Portugal

Barry Mazur Department of Mathematics, Harvard University, Cambridge, MA, USA

Peter Schroeder-Heister Department of Computer Science, University of Tübingen, Tübingen, Germany

Wilfried Sieg Pittsburgh, USA

Craig Smoryński Westmont, IL, USA

Giovanni Sommaruga Eidgenössische Technische Hochschule, Zurich, Switzerland

Chapter 1
Axiomatisches Denken

David Hilbert

Abstract Address delivered by David Hilbert at the annual meeting of the Swiss Mathematical Society in Zurich on September 11, 1917. The German text is from *Mathematische Annalen*, 78:405–415, 1918; the English translation of Joong Fang was first published in Fang, J., editor, *HILBERT–Towards a Philosophy of Modern Mathematics II*, Paideia Press, Hauppauge, N.Y. 1970.

Photography of the participants of the annual meeting of the Swiss Mathematical Society 1917, taken at the back side of the *Landesmuseum* in Zurich. Hilbert is standing in the center of the first row. *Source* Bildarchiv of the ETH Zurich. doi:10.3932/ethz-a-000046430.

D. Hilbert (1862–1943)
Göttingen, Germany

F. Ferreira et al. (eds.), *Axiomatic Thinking I*,
https://doi.org/10.1007/978-3-030-77657-2_1

1.1 Axiomatisches Denken

Wie im Leben der Völker das einzelne Volk nur dann gedeihen kann, wenn es auch allen Nachbarvölkern gut geht, und wie das Interesse der Staaten es erheischt, daß nicht nur innerhalb jedes einzelnen Staates Ordnung herrsche, sondern auch die Beziehungen der Staaten unter sich gut geordnet werden müssen, so ist es auch im Leben der Wissenschaften. In richtiger Erkenntnis dessen haben die bedeutendsten Träger des mathematischen Gedankens stets großes Interesse an den Gesetzen und der Ordnung in den Nachbarwissenschaften bewiesen und vor allem zugunsten der Mathematik selbst von jeher die Beziehungen zu den Nachbarwissenschaften, insbesondere zu den großen Reichen der Physik und der Erkenntnistheorie gepflegt. Das Wesen dieser Beziehungen und der Grund ihrer Fruchtbarkeit, glaube ich, wird am besten deutlich, wenn ich Ihnen diejenige allgemeine Forschungsmethode schildere, die in der neueren Mathematik mehr und mehr zur Geltung zu kommen scheint: ich meine die *axiomatische Methode*.

Wenn wir die Tatsachen eines bestimmten mehr oder minder umfassendenbreak Wissensgebietes zusammenstellen, so bemerken wir bald, daß diese Tatsachen einer Ordnung fähig sind. Diese Ordnung erfolgt jedesmal mit Hilfe eines gewissen *Fachwerkes von Begriffen* in der Weise, daß dem einzelnen Gegenstande des Wissensgebietes ein Begriff dieses Fachwerkes und jeder Tatsache innerhalb des Wissensgebietes eine logische Beziehung zwischen den Begriffen entspricht. Das Fachwerk der Begriffe ist nichts anderes als die *Theorie* des Wissensgebietes.

So ordnen sich die geometrischen Tatsachen zu einer Geometrie, die arithmetischen Tatsachen zu einer Zahlentheorie, die statischen, mechanischen, elektrodynamischen Tatsachen zu einer Theorie der Statik, Mechanik, Elektrodynamik oder die Tatsachen aus der Physik der Gase zu einer Gastheorie. Ebenso ist es mit den Wissensgebieten der Thermodynamik, der geometrischen Optik, der elementaren Strahlungstheorie, der Wärmeleitung oder auch mit der Wahrscheinlichkeitsrechnung und der Mengenlehre. Ja es gilt von speziellen rein mathematischen Wissensgebieten, wie Flächentheorie, Galoisscher Gleichungstheorie, Theorie der Primzahlen nicht weniger als für manche der Mathematik fern liegende Wissensgebiete wie gewisse Abschnitte der Psychophysik oder die Theorie des Geldes.

Wenn wir eine bestimmte Theorie näher betrachten, so erkennen wir allemal, daß der Konstruktion des Fachwerkes von Begriffen einige wenige ausgezeichnete Sätze des Wissensgebietes zugrunde liegen und diese dann allein ausreichen, um aus ihnen nach logischen Prinzipien das ganze Fachwerk aufzubauen.

So genügt in der Geometrie der Satz von der Linearität der Gleichung der Ebene und von der orthogonalen Transformation der Punktkoordinaten vollständig, um die ganze ausgedehnte Wissenschaft der Euklidischen Raumgeometrie allein durch die Mittel der Analysis zu gewinnen. Zum Aufbau der Zahlentheorie ferner reichen die Rechnungsgesetze und Regeln für ganze Zahlen aus. In der Statik übernimmt die gleiche Rolle der Satz vom Parallelogramm der Kräfte, in der Mechanik etwa die

1.2 Axiomatic Thinking

Each nation can do well, as in the life of individuals, only if things go equally well in all of her neighboring nations; the life of sciences is similar to that of states whose interest demands that everything be in order within individual states as properly as their relations be in good order among themselves. Understanding this correctly, the most important carriers of mathematical thoughts have always shown great interest in the law and order in neighboring sciences and, above all for the benefit of mathematics, have cultivated the relations to the neighboring sciences, in particular to physics and epistemology. The essence of these relations and the ground of their fertility will be made most distinct, I believe, if I sketch to you that general method of inquiry which appears to grow more and more significant in modern mathematics; the *axiomatic method*, I mean.

If we collate the facts of a specific field of more or less comprehensive knowledge, then we shortly observe that these facts can be set in order. This order occurs invariably with the aid of certain *framework of concepts* such that there exists correspondence between the individual objects in the field of knowledge and a concept of this framework and between those facts within the field of knowledge and a logical relation among concepts. The framework of concepts is nothing but the *theory* of the field of knowledge.

The geometric facts thus order themselves into a geometry, the arithmetic facts into a number theory, the static, mechanic, electrodynamic facts into a theory of statics, mechanics, electrodynamics, or the facts out of physics of gases into a gas theory. The same holds for the fields of knowledge of thermodynamics, of geometric optics, of elementary radiation theory, of heat conduct, or even for probability theory and set theory. Indeed, it holds as well for such specific fields of pure mathematical knowledge as theory of surfaces, Galois theory of equations, theory of primes, and even for none other than some fields of knowledge remotely related to mathematics such as certain segments of psychophysics or economics.

If we examine a specific theory more closely, we then discern on all occasions that at the bottom of the construction of a framework of concepts are certain few prominent propositions of the field of knowledge, which alone are then sufficient for building up the entire framework upon them in accordance with logical principles.

The proposition of the linearity of the planar equation is thus sufficient in geometry, and that of orthogonal transformation of point-coordinates is complete to produce the entirety of extensive knowledge in the geometry of Euclidean space solely by the means of analysis. Similarly, the laws and rules of computation for integers are sufficient for setting up the theory of numbers. The same role is taken over by the parallelogram law of forces in statics, something like Lagrangian

Lagrangeschen Differentialgleichungen der Bewegung und in der Elektrodynamik die Maxwellschen Gleichungen mit Hinzunahme der Forderung der Starrheit und Ladung des Elektrons. Die Thermodynamik läßt sich vollständig auf den Begriff der Energiefunktion und die Definition von Temperatur und Druck als Ableitungen nach ihren Variabeln, Entropie und Volumen, aufbauen. Im Mittelpunkt der elementaren Strahlungstheorie steht der Kirchhoffsche Satz über die Beziehungen zwischen Emission und Absorption; in der Wahrscheinlichkeitsrechnung ist das Gaußsche Fehlergesetz, in der Gastheorie der Satz von der Entropie als negativem Logarithmus der Wahrscheinlichkeit des Zustandes, in der Flächentheorie die Darstellung des Bogenelementes durch die quadratische Differentialform, in der Gleichungstheorie der Satz von der Wurzelexistenz, in der Theorie der Primzahlen der Satz von der Realität und Häufigkeit der Nullstellen der Riemannschen Funktion $\zeta(s)$ der grundlegende Satz.

Diese grundlegenden Sätze können von einem ersten Standpunkte aus als die *Axiome der einzelnen Wissensgebiete* angesehen werden: die fortschreitende Entwicklung des einzelnen Wissensgebietes beruht dann lediglich in dem weiteren logischen Ausbau des schon aufgeführten Fachwerkes der Begriffe. Zumal in der reinen Mathematik ist dieser Standpunkt der vorherrschende, und der entsprechenden Arbeitsweise verdanken wir die mächtige Entwicklung der Geometrie, der Arithmetik, der Funktionentheorie und der gesamten Analysis.

Somit hatte dann in den genannten Fällen das Problem der Begründung der einzelnen Wissensgebiete eine Lösung gefunden; diese Lösung war aber nur eine vorläufige. In der Tat machte sich in den einzelnen Wissensgebieten das Bedürfnis geltend, die genannten, als Axiome angesehenen und zugrunde gelegten Sätze selbst zu begründen. So gelangte man zu "Beweisen" für die Linearität der Gleichung der Ebene und die Orthogonalität der eine Bewegung ausdrückenden Transformation, ferner für die arithmetischen Rechnungsgesetze, für das Parallelogramm der Kräfte, für die Lagrangeschen Bewegungsgleichungen und das Kirchhoffsche Gesetz über Emission und Absorption, für den Entropiesatz und den Satz von der Existenz der Wurzeln einer Gleichung.

Aber die kritische Prüfung dieser "Beweise" läßt erkennen, daß sie nicht an sich Beweise sind, sondern im Grunde nur die Zurückführung aufgewisse tiefer liegende Sätze ermöglichen, die nunmehr ihrerseits an Stelle der zu beweisenden Sätze als neue Axiome anzusehen sind. So entstanden die eigentlichen heute sogenannten *Axiome* der Geometrie, der Arithmetik, der Statik, der Mechanik, der Strahlungstheorie oder der Thermodynamik. Diese Axiome bilden eine tiefer liegende Schicht von Axiomen gegenüber derjenigen Axiomschicht, wie sie durch die vorhin genannten zuerst zugrundegelegten Sätze in den einzelnen Wissensgebieten charakterisiert worden ist. Das Verfahren der axiomatischen Methode, wie es hierin ausgesprochen liegt, kommt also einer *Tieferlegung der Fundamente* der einzelnen Wissensgebiete gleich, wie eine solche ja bei jedem Gebäude nötig wird in dem Maße, als man dasselbe ausbaut, höher führt und dennoch für seine Sicherheit bürgen will.

differential equations of motion in mechanics, and Maxwell's equations accepting the condition of rigidity and change of electrons in electrodynamics. Thermodynamics is completely built upon the concept of energy function and the definition of temperature and pressure as derivatives from their variables, entropy, and volume. In the midpoint of elementary radiation theory there stands Kirchhoff's law on the relations between emission and absorption; there is the Gaussian law of error in the calculus of probability, the theorem of entropy as negative logarithm of probability of events in the gas theory, the representation of arc elements by quadratic differential form in the theory of surfaces, the existence theorem of roots in the theory of equations, the reality and frequency theorem of zero-points of Riemann zeta function, the fundamental theorem in the theory of primes.

Viewed from a primary standpoint, these theorems may be looked upon as the *axioms of individual fields of knowledge*; the advancing development of individual fields of knowledge rests then on the more extensive logical enlargement of the completed framework of concepts. This standpoint predominates principally in pure mathematics, and we are indebted for the corresponding modes of operation to the mighty development of geometry, arithmetic, function theory, and analysis in entirety.

In the preceding cases the problem of founding individual fields of knowledge had consequently obtained a solution; this solution, however, was only a tentative one. As a matter of fact, it became necessary in individual fields of knowledge to found anew the aforementioned founding propositions themselves, once considered axioms and placed at the foundation. So came into being the "proofs" for the linearity of the planar equation and the orthogonality of the transformation expressing a motion, and also for the laws of arithmetic computations, for the parallelogram of forces, for the Lagrangian equations of motion and Kirchhoff's law of emission and absorption, for the principle of entropy and the existence theorem of the roots of an equation.

But the critical examination of these "proofs" makes it discernible that they are not proofs in themselves; rather, they are in the main only capable of leading back to certain more deeply lying propositions which in turn are now to be looked upon as new axioms in place of those propositions to be proved. So originated the proper, currently so-called axioms of geometry, of arithmetic, of statics, of mechanics, of radiation theory or of thermodynamics. These axioms form a more deeply lying layer of axioms opposite to another layer of axioms as they have been characterized by the aforementioned propositions of the first foundation in individual fields of knowledge. The procedure of the axiomatic method, as articulated here, comes thus up to a *deeper-laying of foundations* of individual fields of knowledge, just as such a one is indeed necessary for each building according as it will be enlarged, built higher, and yet vouched for its safety.

Soll die Theorie eines Wissensgebietes, d. h. das sie darstellende Fachwerk der Begriffe, ihrem Zwecke, nämlich der Orientierung und Ordnung dienen, so muß es vornehmlich gewissen zwei Anforderungen genügen: *erstens* soll es einen Überblick über die *Abhängigkeit* bzw. *Unabhängigkeit* der Sätze der Theorie und *zweitens* eine Gewähr der *Widerspruchslosigkeit* aller Sätze der Theorie bieten. Insbesondere sind die Axiome einer jedenTheorie nach diesen beiden Gesichtspunkten zu prüfen.

Beschäftigen wir uns zunächst mit der Abhängigkeit bzw. Unabhängigkeit der Axiome.

Das klassische Beispiel für die Prüfung der Unabhängigkeit eines Axioms bietet das *Parallelenaxiom* in der Geometrie. Die Frage, ob der Parallelensatz durch die anderen Axiome schon bedingt ist, verneinte Euklid, indem er ihn unter die Axiome setzte. Die Untersuchungsmethode Euklids wurde vorbildlich für die axiomatische Forschung, und seit Euklid ist zugleich die Geometrie das Musterbeispiel für eine axiomatisierte Wissenschaft überhaupt.

Ein anderes Beispiel für eine Untersuchung über die Abhängigkeit der Axiome bietet die klassische Mechanik. Vorläufigerweise konnten, wie vorhin bemerkt, die Lagrangeschen Gleichungen der Bewegung als Axiome der Mechanik gelten — läßt sich doch auf diese in ihrer allgemeinen Formulierung für beliebige Kräfte und beliebige Nebenbedingungen die Mechanik gewiß vollständig gründen. Bei näherer Untersuchung zeigt sich aber, daß beim Aufbau der Mechanik sowohl beliebige Kräfte wie beliebige Nebenbedingungen vorauszusetzen unnötig ist und somit das System von Voraussetzungen vermindert werden kann. Diese Erkenntnis führt einerseits zu dem Axiomensystem von BOLTZMANN, der nur Kräfte, und zwar speziell Zentralkräfte, aber keine Nebenbedingungen annimmt und dem Axiomensystem von HERTZ, der die Kräfte verwirft und mit Nebenbedingungen, und zwar speziell mit festen Verbindungen auskommt. Diese beiden Axiomensystemebilden somit eine tiefere Schicht in der fortschreitenden Axiomatisierung der Mechanik.

Nehmen wir bei Begründung der Galoisschen Gleichungstheorie die Existenz der Wurzeln einer Gleichung als Axiom an, so ist dieses sicher ein abhängiges Axiom; denn jener Existenzsatz ist aus den arithmetischen Axiomen beweisbar, wie zuerst GAUSS gezeigt hat.

Ähnlich verhält es sich damit, wenn wir etwa den Satz von der Realität der Nullstellen der Riemannschen Funktion $\zeta(s)$ in der Primzahlentheorie als Axiom annehmen wollten: beim Fortschreiten zur tieferen Schicht der reinen arithmeti- schen Axiome würde der Beweis dieses Realitätssatzes notwendig sein und dieser erst uns die Sicherheit der wichtigen Folgerungen gewähren, die wir durch seine Postulierung schon jetzt für die Theorie der Primzahlen aufgestellt haben.

Besonderes Interesse für die axiomatische Behandlung bietet die Frage der Abhängigkeit der Sätze eines Wissensgebietes von dem Axiom der *Stetigkeit.*

In der Theorie der reellen Zahlen wird gezeigt, daß das Axiom des Messens, das sogenannte Archimedische Axiom, von allen übrigen arithmetischen Axiomen unabhängig ist. Diese Erkenntnis ist bekanntlich für die Geometrie von wesentlicher Bedeutung, scheint mir aber auch für die Physik von prinzipiellem Interesse;

If the theory of a field of knowledge, that is, the framework of concepts that represents the theory, is to serve its purpose, namely the orientation and order, it must then satisfy chiefly two fixed demands: it must offer, first, a general view of the *dependence* or *independence* of the propositions of the theory and, second, a guarantee of *consistency* of all propositions of the theory. In particular, the axioms of each theory have to be proved in accordance with these two viewpoints.

Let us work first at the dependence or independence of axioms.

The *parallel axiom* in geometry offered the classic example for the examination of independence of an axiom. Euclid answered in the negative to the question as to whether the proposition of parallels is already conditioned by other axioms, because he placed it under the axioms. Euclid's method of investigation became typical of the axiomatic investigation and, since Euclid, geometry has at once been the model example for an axiomatic science in general.

Classic mechanics offers another example for an investigation of independence of axioms. The Lagrangian equation of motion, as has already been observed, could act as axioms of mechanics—upon these does mechanics no doubt found itself completely in their general formulation for arbitrary forces and arbitrary secondary conditions. A closer examination reveals, however, that arbitrary forces as well as arbitrary secondary conditions are unnecessary to presuppose for the construction of mechanics, and that, consequently, the system of presuppositions can be reduced. This recognition leads on the one hand to the axiomatic system of Boltzmann who presupposes only forces, indeed central forces in particular, and on the other hand to the axiomatic system of Hertz who rejects forces and wants no more than the secondary conditions, indeed fixed connections in particular. These two axiomatic systems form thus a deeper layer in the advancing axiomatization of mechanics.

It is similarly the case if we would assume as an axiom something like the reality theorem of zero points of Riemann zeta function in the theory of primes. The proof of this reality theorem would become necessary for the progress towards the deeper layer of pure arithmetic axioms, and it would best guarantee the safety of important conclusions; we have already set up the axioms through its postulation for the theory of primes.

Special interest for the axiomatic treatment is offered by the question of dependence of the propositions of a field of knowledge on the axiom of continuity.

It is shown in the theory of real numbers that the axiom of measurement, the so-called Archimedian axiom, is independent of all other axioms of arithmetic. As is well-known, this knowledge is of essential significance to geometry, but it seems to me that it has principal interest in physics as well; for it leads us to the following

denn sie führt uns zu folgendem Ergebnis: die Tatsache,daß wir durch Aneinanderfügen irdischer Entfernungen die Dimensionen und Entfernungen der Körper im Weltenraume erreichen, d. h. durch irdisches Maß die himmlischen Längen messen können, ebenso die Tatsache, daß sich die Distanzen im Atominneren durch das Metermaß ausdrücken lassen, sind keineswegs bloß eine logische Folge der Sätze über Dreieckskongruenzen und der geometrischen Konfiguration, sondern erst ein Forschungsresultat der Empirie. Die Gültigkeit des Archimedischen Axioms in der Natur bedarf eben im bezeichneten Sinne gerade so der Bestätigung durch das Experiment wie etwa der Satz von der Winkelsumme im Dreieck im bekannten Sinne.

Allgemein möchte ich das Stetigkeitsaxiom in der Physik wie folgt formulieren: "Wird für die Gültigkeit einer physikalischen Aussage irgend ein beliebiger Genauigkeitsgrad vorgeschrieben, so lassen sich kleine Bereiche angeben, innerhalb derer die für die Ausage gemachten Voraussetzungen frei variieren dürfen, ohne daß die Abweichung von der Aussage den vorgeschriebenen Genauigkeitsgrad überschreitet." Dies Axiom bringt im Grunde nur zum Ausdruck, was unmittelbar im Wesen des Experimentes liegt; es ist stets von den Physikern angenommen worden, ohne daß es bisher besonders formuliert worden ist.

Wenn man z. B. nach PLANCK aus dem Axiom der Unmöglichkeit des *Perpetuum mobile zweiter Art* den zweiten Wärmesatz ableitet, so wird dabei dieses Stetigkeitsaxiom notwendigerweise benutzt.

Daß in der Begründung der Statik beim Beweise des Satzes vom *Parallelogramm der Kräfte* das Stetigkeitsaxiom notwendig ist — wenigstens bei einer gewissen nächstliegenden Auswahl der übrigen Axiome — hat HAMEL auf eine sehr interessante Weise durch Heranziehung des Satzes von der Wohlordnungsfähigkeit des Kontinuums gezeigt.

Die Axiome der klassischen Mechanik können eine Tieferlegung erfahren, wenn man sich vermöge des Stetigkeitsaxioms die kontinuierliche Bewegung in kurz aufeinanderfolgende, geradlinig gleichförmige stückweise durch Impulse hervorgerufene Bewegungen zerlegt denkt und dann als wesentliches mechanisches Axiom das *Bertrandsche Maximalprinzip* verwendet, demzufolge nach jedem Stoß die wirklich eintretende Bewegung stets diejenige ist, bei welcher die kinetische Energie des Systems ein Maximum wird gegenüber allen mit dem Satz von der Erhaltung der Energie verträglichen Bewegungen.

Auf die neuesten Begründungsarten der Physik, insbesondere der Elektrodynamik, die ganz und gar Kontinuumstheorien sind und dem gemäß die Stetigkeitsforderung in weitestem Maße erheben, möchte ich hier nicht eingehen, weil diese Forschungen noch nicht genügend abgeschlossen sind.

Wir wollen nun den zweiten der vorhin genannten Gesichtspunkte, nämlich die Frage nach der *Widerspruchslosigkeit* der Axiome prüfen; diese ist offenbar von höchster Wichtigkeit, weil das Vorhandensein eines Widerspruches in einer Theorie offenbar den Bestand der ganzen Theorie gefährdet.

outcome. That is, the fact that we can come up with the dimensions and ranges of celestial bodies by putting together terrestial ranges, namely measuring celestial lengths by terrestial measure, as well as the fact that the distances inside atoms can be expressed in terms of metric measure, is by no means a merely logical consequence of propositions on the triangular congruence and the geometric configuration, but rather an investigative result of experience. The validity of the Archimedian axiom in the nature, in the sense indicated as above, needs experimental confirmation just as much as does the proposition of the angular sum in triangle in the ordinary sense.

In general, I should like to formulate the axiom of continuity in physics as follows: "If a certain arbitrary degree of exactitude is prescribed for the validity of a physical assertion, a small range shall then be specified, within which the presuppositions prepared for the assertion may freely vary such that the deviation from the assertion does not overstep the prescribed degree of exactitude." This axiom in the main brings only that into expression which directly lies in the essense of experiments; it has always been assumed by physicists who, however, have never specifically formulated it.

If one derives, after Planck for instance, the second Heat Theorem from the axiom of the impossibility of the *Perpetuum mobile zweiter Art* (perpetual motion machine of the second kind), this axiom of continuity is then necessarily employed for it.

Hamel has shown in a very interesting manner, referring to the well-ordering principle of the continuum, that the axiom of continuity is necessary for the proof of the parallel law of forces in the foundation of statics—at least for a certain handy choice of other axioms.

The axioms of classic mechanics can experience a process of deeper founding if one considers the continuous motion, by virtue of the axiom of continuity, in short successive, rectlinearly uniform piecewise broken motions caused by impulse, and then employs Bertrand's maximal principle as a primary mechanical axiom; according to the latter, the actually occurring motion after each push is what makes the kinetic energy of the system a maximum opposite to all motions compatible with the principle of the conservation of energy.

Into the newest modes of founding physics, especially of electrodynamics, which in entirety is nothing but the theory of continuum itself and accordingly takes up the challenge of continuity to the widest extent, I would not enter here, since this investigation has not yet been sufficiently completed.

We will now examine the second of the aforementioned viewpoints, namely the question of the consistency of axioms; this is manifestly of greater importance, since the presence of a contradiction in a theory manifestly imperils the stability of the entire theory.

Die Erkenntnis der inneren Widerspruchslosigkeit ist selbst bei längst anerkannten und erfolgreichen Theorien mit Schwierigkeit verbunden: ich erinnere an den *Umkehr- und Wiederkehreinwand* in der kinetischen Gastheorie.

Oftmals passiert es, daß die innere Widerspruchslosigkeit einer Theorie als selbstverständlich angesehen wird, während in Wahrheit tiefe mathematische Entwicklungen zu dem Nachweise nötig sind. Als Beispiel betrachten wir ein Problem aus der elementaren Theorie der *Wärmeleitung*, nämlich die Temperaturverteilung innerhalb eines homogenen Körpers, dessen Oberfläche auf einer bestimmten von Ort zu Ort variierenden Temperatur gehalten wird: alsdann enthält in der Tat die Forderung des Bestehens von Temperaturgleichgewicht keinen inneren Widerspruch der Theorie. Zur Erkenntnis dessen ist aber der Nachweis nötig, daß die bekannte Randwertaufgabe der Potentialtheorie stets lösbar ist; denn erst die Lösung dieser Randwertaufgabe zeigt, daß eine der Wärmeleitungsgleichung genügende Temperaturverteilung überhaupt möglich ist.

Aber zumal in der Physik genügt es nicht, wenn die Sätze einer Theorie unter sich in Einklang stehen; vielmehr ist noch die Forderung zu erheben, daß sie auch den Sätzen eines benachbarten Wissensgebietes niemals widersprechen.

So liefern, wie ich kürzlich zeigte, die Axiome der elementaren Strahlungstheorie außer der Begründung des *Kirchhoffschen Satzes* über Emission und Absorption noch einen speziellen Satz über Reflexion und Brechung einzelner Lichtstrahlen, nämlich den Satz: Wenn zwei Strahlen natürlichen Lichtes und gleicher Energie von je einer Seite her auf die Trennungsfläche zweier Medien in solchen Richtungen auffallen, daß der eine Strahl nach seinem Durchtritt, der andere nach seiner Reflexion dieselbe Richtung aufweist, so ist der durch die Vereinigung entstehende Strahl wieder von natürlichem Licht und gleicher Energie. Dieser Satz ist — wie sich in der Tat zeigt — mit der Optik keineswegs in Widerspruch, sondern kann als Folgerung aus der elektromagnetischen Lichttheorie abgeleitet werden.

Die Resultate der *kinetischen Gastheorie* stehen bekanntlich mit der *Thermodynamik* in bestem Einklang.

Ebenso sind *elektromagnetische Trägheit* und *Einsteinsche Gravitation* mit den entsprechenden Begriffen der klassischen Theorien verträglich, insofern diese letzteren als Grenzfälle der allgemeineren Begriffe in den neuen Theorien aufzufassen sind.

Dagegen hat die *moderne Quantentheorie* und die fortschreitende Erkenntnis der inneren Atomstruktur zu Gesetzen geführt, die der bisherigen wesentlich auf den Maxwellschen Gleichungen aufgebauten Elektrodynamik geradezu widersprechen; die heutige Elektrodynamik bedarf daher — wie jedermann anerkennt — notwendig einer neuen Grundlegung und wesentlichen Umgestaltung.

Wie man aus dem bisher Gesagten ersieht, wird in den physikalischen Theorien die Beseitigung sich einstellender Widersprüche stets durch veränderte Wahl der Axiome erfolgen müssen, und die Schwierigkeit besteht darin, die Auswahl so zu treffen, daß alle beobachteten physikalischen Gesetze logische Folgen der ausgewählten Axiome sind.

The understanding of the internal consistency is linked to difficulty even in the long accepted and flourishing theories; I am thinking of "*Umkehr- und Wiederkehreinwand*" (arguments against Boltzmann's "H-Theorem" by Loschmidt and Zermelo, respectively) in the kinetic gas theory.

It often happens that the internal consistency of a theory is considered self-explanatory while, in truth, deep mathematical developments are necessary for proofs. For example, let us reflect upon a problem from the elementary theory of *heat conduction*, namely the temperature distribution inside a homogeneous body whose surface is kept well within a definite temperature varying from spot to spot; then, the demand for the existence of temperature equilibrium contains in fact no internal contradiction of the theory. To understand this, however, it is necessary to prove that the well-known boundary value problem of potential theory is always solvable, because this boundary value problem alone shows that a temperature distribution satisfying the equation of heat conduction is possible at all.

It is all the more insufficient in physics, however, if the propositions of a theory are in harmony with themselves; rather, they still have to meet the demand that they never contradict the propositions of a neighboring field of knowledge as well.

Thus, as I showed a short while ago, the axioms of elementary radiation theory affords, in addition to the foundation of Kirchhoff's law on emission and absorption, still another special law on reflection and refraction of single light rays, namely the law: If two rays of natural light and of equal energy fall at a time from a side upon the area of separation of two media in such directions that one ray after penetration and the other after reflection take the same direction, the ray created by the union is then again of natural light and of equal energy. This law, as it shows in fact, is in no way contradicting optics, but it can be derived as conclusion from the electromagnetic theory of light.

The results of the kinetic gas theory are well-known to be in the best harmony with thermodynamics.

Similarly, *electromagnetic inertia* and *Einstein's gravitation* are compatible with their corresponding concepts in classic mechanics as long as the latter are considered borderline cases of the more general concepts.

On the other hand, the *modern theory of quanta* and the advancing knowledge of the inner structure of atoms have led to the laws which flatly contradict the electrodynamics that, up to now, has substantially been built upon Maxwell's equations; as everyone concedes, therefore, the contemporary electrodynamics necessarily demands a new foundation and substantial modification.

As one notes from the preceding lines, the elimination of recurring contradictions in physical theories must always take place through altered selection of axioms, and the difficulty lies in the proper choice by which all observed physical laws are logically deducible.

Anders verhält es sich, wenn in rein theoretischen Wissensgebieten Widersprüche auftreten. Das klassische Beispiel für ein solches Vorkommnis bietet die Mengentheorie, und zwar insbesondere das schon auf CANTOR zurückgehende *Paradoxon der Menge aller Mengen*. Dieses Paradoxon ist so schwerwiegend, daß sehr angesehene Mathematiker, z. B. KRONECKER und POINCARÉ, sich durch dasselbe veranlaßt fühlten, der gesamten Mengentheorie — einem der fruchtreichsten und kräftigsten Wissenszweige der Mathematik überhaupt — die Existenzberechtigung abzusprechen.

Auch bei dieser prekären Sachlage brachte die axiomatische Methode Abhilfe. Es gelang ZERMELO, indem er durch Aufstellung geeigneter Axiome einerseits die Willkür der Definitionen von Mengen und andererseits die Zulässigkeit von Aussagen über ihre Elemente in bestimmter Weisebeschränkte, die Mengentheorie derart zu entwickeln, daß die in Rede stehenden Widersprüche wegfallen, daß aber trotz der auferlegten Beschränkungen die Tragweite und Anwendungsfähigkeit der Mengentheorie die gleiche bleibt.

In allen bisherigen Fällen handelte es sich um Widersprüche, die sich im Verlauf der Entwicklung einer Theorie herausgestellt hatten und zu deren Beseitigung durch Umgestaltung des Axiomensystems die Not drängte. Aber es genügt nicht, vorhandene Widersprüche zu vermeiden, wenn der durch sie gefährdete Ruf der Mathematik als Muster strengster Wissenschaft wiederhergestellt werden soll: die prinzipielle Forderung der Axiomenlehre muß vielmehr weitergehen, nämlich dahin, zu erkennen, daß jedesmal innerhalb eines Wissensgebietes auf Grund des aufgestellten Axiomensystems Widersprüche *überhäuft unmöglich* sind.

Dieser Forderung entsprechend habe ich in den *Grundlagen der Geometrie* die Widerspruchslosigkeit der aufgestellten Axiome nachgewiesen, indem ich zeigte, daß jeder Widerspruch in den Folgerungen aus dengeometrischen Axiomen notwendig auch in der Arithmetik des Systems der reellen Zahlen erkennbar sein müßte.

Auch für die physikalischen Wissensgebiete genügt es offenbar stets, die Frage der *inneren Widerspruchslosigkeit* auf die Widerspruchslosigkeitder arithmetischen Axiome zurückzuführen. So zeigte ich die Widerspruchslosigkeit der Axiome *der elementaren Strahlungstheorie*, indem ich für dieselbe das Axiomensystem aus analytisch unabhängigen Stücken aufbaute — die Widerspruchslosigkeit der Analysis dabei voraussetzend.

Ähnlich darf und soll man unter Umständen beim Aufbau einer mathematischen Theorie verfahren. Haben wir beispielsweise bei Entwicklung der Galoisschen Gruppentheorie den Satz von der *Wurzelexistenz* oder in der Theorie der Primzahlen den Satz von der *Realität der Nullstellen* der Riemannschen Funktion $\zeta(s)$ als Axiom betrachtet, so läuft jedesmal der Nachweis der Widerspruchslosigkeit des Axiomensystems eben daraufhinaus, den Satz von der Wurzelexistenz bzw. den Riemannschen Satz über die Funktion $\zeta(s)$ mit den Mitteln der Analysis zu beweisen — und damit erst ist die Vollendung der Theorie gesichert.

Auch die Frage der Widerspruchslosigkeit des Axiomensystems für die *reellen Zahlen* läßt sich, durch Benutzung mengentheoretischer Begriffe, auf die nämliche Frage für die ganzen Zahlen zurückführen: dies ist das Verdienst der Theorien der Irrationalzahlen von WEIERSTRASS und DEDEKIND.

The situation changes, however, if contradictions appear in purely theoretical fields of knowledge. Set theory offers the classic example for such an occurrence, in particular the paradox of the set of all sets that goes as far back as to Cantor himself. This paradox is so serious that such very distinguished mathematicians as Kronecker and Poincaré for instance felt induced to deny set theory in entirety—one of the most fruitful and powerful branches of mathematics in general—any justification of existence.

The axiomatic method brought remedy also under those precarious circumstances. As he set up suitable axioms to restrict, on the one hand, the arbitrariness in the definitions of sets themselves and, on the other, the admissibility of statements on their elements in a specific way, Zermelo succeeded to develop set theory in such a manner that the paradoxes under discussion fall away and, for all restrictions, the purport and applicability of set theory remains the same.

In all the preceding cases the question was the contradictions which had been brought out in the process of developing a theory, and their elimination pressed the need for modifications of axiomatic systems. It is not enough to avoid contradictions, however, if the reputation, damaged by them, of mathematics as the model of the most rigorous science should be restored. The principal demand of axiomatics must go further ahead, namely so far as to understand that contradictions are always *altogether impossible* in a field of knowledge founded on the erected system of axioms.

Corresponding to this demand in the *Grundlagen der Geometrie* (Foundations of Geometry) I proved the consistency of the erected axioms, in which I showed that each contradiction in the deduction from the geometric axioms must necessarily be discernible in the arithmetic of the real number system as well.

No doubt for the fields of physical knowledge, too, it is always sufficient to reduce the question of *inner consistency* to the consistency of arithmetic axioms. So I showed the consistency of the axioms of elementary radiation theory as I constructed the axiomatic system for the theory with analytically independent parts—where the consistency of analysis was presumed.

Circumstances permitting, one may and must similarly proceed in the construction of a mathematical theory. If we regard as an axiom, for example, the theorem of the *existence of roots* in the development of Galois theory of equations or the theorem of the *reality of zero points* of Riemann zeta function in the theory of primes, the consistency proof of the axiomatic system then always amounts just to the proof of the theorem of the existence of roots and Riemann's theorem on the zeta function with the aid of analysis, respectively—and the completion of the theory is therewith secured for the first time.

Also the question of the consistency of the axiomatic system for *real numbers* is reduced, through the use of set theoretic concepts, to the same question for integers. This is the merit of the theories, by Weierstrass and Dedekind, of irrational numbers.

Nur in zwei Fällen nämhch, wenn es sich um die Axiome der *ganzen Zahlen* selbst und wenn es sich um die Begründung der *Mengenlehre* handelt, ist dieser Weg der Zurückführung auf ein anderes spezielleres Wissensgebiet offenbar nicht gangbar, weil es außer der Logik überhaupt keine Disziplin mehr gibt, auf die alsdann eine Berufung möglich wäre.

Da aber die Prüfung der Widerspruchslosigkeit eine unabweisbare Aufgabe ist, so scheint es nötig, die Logik selbst zu axiomatisieren und nachzuweisen, daß Zahlentheorie sowie Mengenlehre nur Teile der Logik sind.

Dieser Weg, seit langem vorbereitet — nicht zum mindesten durch die tiefgehenden Untersuchungen von FREGE — ist schließlich am erfolgreichsten durch den scharfsinnigen Mathematiker und Logiker RUSSELL eingeschlagen worden. In der Vollendung dieses großzügigen Russellschen Unternehmens der *Axiomatisierung der Logik* könnte man die Krönung des Werkes der Axiomatisierung überhaupt erblicken.

Diese Vollendung wird indessen noch neuer und vielseitiger Arbeit bedürfen. Bei näherer Überlegung erkennen wir nämlich bald, daß die Frage der Widerspruchslosigkeit bei den ganzen Zahlen und Mengen nicht eine für sich alleinstehende ist, sondern einem großen Bereiche schwierigster erkenntnistheoretischer Fragen von spezifisch mathematischer Färbung angehört: ich nenne, um diesen Bereich von Fragen kurz zucharakterisieren, das Problem der prinzipiellen *Lösbarkeit einer jeden mathematischen Frage*, das Problem der nachträglichen *Kontrollierbarkeit* des Resultates einer mathematischen Untersuchung, ferner die Frage nach einem *Kriterium für die Einfachheit* von mathematischen Beweisen, die Frage nach dem Verhältnis zwischen *Inhaltlichkeit* und *Formalismus* in Mathematik und Logik und endlich das Problem der *Entscheidbarkeit* einer mathematischen Frage durch eine endliche Anzahl von Operationen.

Wir können uns nun nicht eher mit der Axiomatisierung der Logik zufrieden geben, als bis alle Fragen dieser Art in ihrem Zusammenhange verstanden und aufgeklärt sind.

Unter den genannten Fragen ist die letzte, nämlich die Frage nach der Entscheidbarkeit durch eine endliche Anzahl von Operationen, die bekannteste und die am häufigsten diskutierte, weil sie das Wesen des mathematischen Denkens tief berührt.

Ich möchte das Interesse für sie zu vermehren suchen, indem ich auf einige speziellere mathematische Probleme hinweise, in denen sie eine Rolle spielt.

In der Theorie der *algebraischen Invarianten* gilt bekanntlich der Fundamentalsatz, daß es stets eine endliche Anzahl von ganzen rationalen Invarianten gibt, durch die sich alle übrigen solchen Invarianten in ganzer rationaler Weise darstellen lassen. Der erste von mir angegebene allgemeine Beweis für diesen Satz befriedigt, wie ich glaube, unsere Ansprüche, was Einfachheit und Durchsichtigkeit anlangt, vollauf; es ist aber unmöglich, diesen Beweis so umzugestalten, daß wir durch ihn eine angebbare Grenze für die Anzahl der endlich vielen Invarianten des vollen Systems erhalten oder gar zur wirklichen Aufstellung derselben gelangen. Es sind vielmehr ganz anders geartete Überlegungen und neue Prinzipien notwendig gewesen, um zu erkennen, daß die Aufstellung des vollen Invariantensystems lediglich Operationen erfordert, deren Anzahl endlich ist und unterhalb einer vor der Rechnung angebbaren Grenze liegt.

Only in two cases, namely if it is a question of the axioms of integers themselves, and if it is a question of the foundation of set theory, this mode of reduction to another specific field of knowledge is manifestly impracticable, since beyond logic there is no more discipline to which an appeal could be lodged.

Since the consistency proof is a task that cannot be dismissed, however, it seems necessary to axiomatize logic itself and then to establish that number theory as well as set theory is only a part of logic.

This road, prepared for a long time—not in the least through the profound investigations by Frege—has finally been pursued by the ingenious mathematician and logician, Russell, with greatest success. In the completion of this extensive enterprise by Russell for the axiomatization of logic one can behold the crowning of the work of axiomatization in general.

Meanwhile, this completion still demands new and various works. By closer examination we presently discern that the consistency question of integers and sets is not isolated for itself; rather, it belongs to a great region of the most difficult epistemological questions of specific mathematical coloration. To characterize briefly this region of questions, I mention by name the problem of principal *solvability of every mathematical question*, the problem of supplementary *controllability* of the results of a mathematical investigation, the question of a *criterion for the simplicity* of mathematical proofs, the question of relations between *contentualness* (*Inhaltlichkeit*) and *formalism* in mathematics and logic, and finally the problem of *decidability* of a mathematical question by a finite number of operations.

Until all questions of this type in their correlation are understood and explained, then, we cannot be satisfied with the axiomatization of logic.

The last among the preceding questions, namely the question of decidability by a finite number of operations, is the most well-known and also the most frequently discussed, because it deeply touches the essence of mathematical thinking.

I should like to stir up more interest in it as I now refer to a few particular mathematical problems in which it plays a role.

In the theory of algebraic invariants, the fundamental theorem is known to hold that there always exists a finite number in (rational) integers of invariants, by which all the rest of such invariants can be integrally represented. The first general proof, by myself, of this theorem satisfies our demands, as I believe, and indeed abundantly with respect to simplicity and lucidity. It is impossible, however, to modify this proof so that we obtain in the process a specifiable limit for the number of finitely many invariants of the full system, or even to succeed in arranging it in concrete. Rather, entirely different kinds of investigation and new principles are necessarily needed for discerning that the arrangement of the full invariant system demands only those operations, the number of which is finite and lies under a limit specifiable by computation.

Das gleiche Vorkommnis bemerken wir an einem Beispiel aus der *Flächentheorie*. In der Geometrie der Flächen vierter Ordnung ist es eine fundamentale Frage, aus wie vielen von einander getrennten Mänteln eine solche Fläche höchstens bestehen kann.

Das erste bei der Beantwortung dieser Frage ist der Nachweis, daß die Anzahl der Flächenmäntel endlich sein muß; dieser kann leicht auf funktionentheoretischem Wege, wie folgt, geschehen. Man nehme das Vorhandensein unendlich vieler Mäntel an und wähle dann innerhalb eines jeden durch einen Mantel begrenzten Raumteiles je einen Punkt aus. Eine Verdichtungsstelle dieser unendlich vielen ausgewählten Punkte würde dann ein Punkt von solcher Singularität sein, wie sie für eine algebraische Fläche ausgeschlossen ist.

Dieser funktionentheoretische Weg führt auf keine Weise zu einer oberen Grenze für die Anzahl der Flächenmäntel; dazu bedarf es vielmehr gewisser Überlegungen, über Schnittpunktsanzahlen, die dann schließlich lehren, daß die Anzahl der Mäntel gewiß nicht größer als 12 sein kann.

Die zweite von der ersten gänzlich verschiedene Methode läßt sich ihrerseits nicht dazu anwenden und auch nicht so umgestalten, daß sie die Entscheidung ermöglicht, ob eine Fläche 4. Ordnung mit 12 Mänteln wirklich existiert.

Da eine quaternäre Form 4. Ordnung 35 homogene Koeffizienten besitzt, so können wir uns eine bestimmte Fläche 4. Ordnung durch einen Punkt im 34-dimensionalen Raume veranschaulichen. Die Diskriminante der quaternären Form 4. Ordnung ist vom Grade 108 in den Koeffizienten derselben; gleich Null gesetzt, stellt sie demnach im 34-dimensionalen Räume eine Fläche 108. Ordnung dar. Da die Koeffizienten der Diskriminante selbst bestimmte ganze Zahlen sind, so läßt sich der topologische Charakter der Diskriminantenfläche nach den Regeln, die uns für den 2- und 3-dimensionalen Raum geläufig sind, genau feststellen, so daß wir über die Natur und Bedeutung der einzelnen Teilgebiete, in die die Diskriminantenfläche den 34-dimensionalen Raum zerlegt, genaue Auskunft erhalten können. Nun besitzen die durch Punkte des nämlichen Teilgebietes dargestellten Flächen 4. Ordnung gewiß alle die gleiche Mäntelzahl, und es ist daher möglich, durch eine endliche, wenn auch sehr mühsame und langwierige Rechnung, festzustellen, ob eine Fläche 4. Ordnung mit $n \leq 12$ Mänteln vorhanden ist oder nicht.

Die eben angestellte geometrische Betrachtung ist also ein dritter Weg zur Behandlung unserer Frage nach der Höchstzahl der Mäntel einer Fläche 4. Ordnung. Sie beweist die Entscheidbarkeit dieser Frage durch eine endliche Anzahl von Operationen. Prinzipiell ist damit eine bedeutende Förderung unseres Problems erreicht: dasselbe ist zurückgeführt auf ein Problem von dem Range etwa der Aufgabe, die $10^{(10^{10})}$-te Ziffer der Dezimalbruchentwicklung von π zu ermitteln — einer Aufgabe, deren Lösbarkeit offenbar ist, deren Lösung aber unbekannt bleibt.

Vielmehr bedurfte es einer von ROHN ausgeführten tiefgehenden schwierigen algebraisch-geometrischen Untersuchung, um einzusehen, daß bei einer Fläche 4. Ordnung 11 Mäntel nicht möglich sind; 10 Mäntel dagegen kommen wirklich vor. Erst diese vierte Methode bringt somit die völlige Lösung des Problems.

The same situation is observed in an example from the theory of surfaces. In the geometry of fourth-order surfaces there is a fundamental question as to how many of mutually separated convex surfaces (i.e. sheets) such a surface at most consists of.

The first to be answered for the question is the evidence that the number of such sheets must be finite; this can easily appear on the road of function theory, namely as follows: Assume the presence of infinitely many sheets and then choose a point at a time inside a part of space bounded by a sheet. A concentration spot of these infinitely many chosen points would then be a point of such singularity as to be barred from an algebraic surface.

This road of function theory leads by no means to an upper limit for the number of sheets; for this, one should rather investigate the number of intersections, which will then finally show that the number of sheets can never be more than 12.

The second method, completely different from the first, in turn is not applicable to it and also not modifiable so that it will enable us to decide whether there exists a fourth-order surface with 12 sheets.

Since a fourth-order quaternary form has 35 homegeneous coefficients, we can then illustrate a specific fourth-order surface through a point in a 34-dimensional space. The discriminant of the fourth-order quaternary form is of degree 108 in its own coefficients; if equated to zero, it then represents a 108th-order surface in the 34-dimensional space. Since the coefficients of the discriminant itself are specific integers, the topological character of the discriminant surface is exactly determinable according to the rules which are routine in the 2- and 3-dimensional spaces such that we can be exactly informed of the nature and meaning of individual sections in which the discriminant surface divides the 34-dimensional space. Now all the fourth-order surfaces represented by the points of these sections certainly possess the equal number of sheets, and it is therefore possible to determine, by finite if only very troublesome and wearisome steps of computation, whether or not there exists a fourth-order surface with less than or equal to 12 sheets.

The geometric consideration as above is thus a third way to treat our question concerning the maximal number of sheets of a fourth-order surface. It proves the decidability of this question by a finite number of operations. An important demand of our problem is thereby met in principle. Similarly reduced to a problem of the same rank is something like the task of establishing the decimal expression of π up to the $10^{10^{10}}$th place—a task, the solvability of which is evident, but the solution of which remains unknown.

It needs perhaps the penetrating and difficult investigation in algebraic geometry, carried out by Rohn, to understand that Jl sheets are impossible for a fourth-order surface; on the other hand, 10 sheets actually occur. This fourth method alone brings therewith the complete solution of the problem.

Diese speziellen Ausführungen zeigen, wie verschiedenartige Beweismethoden auf dasselbe Problem anwendbar sind, und sollen nahelegen, wie notwendig es ist, das Wesen des mathematischen Beweises an sich zu studieren, wenn man solche Fragen, wie die nach der Entscheidbarkeit durch endlich viele Operationen, mit Erfolg aufklären will.

Alle solchen prinzipiellen Fragen, wie ich sie vorhin charakterisierte und unter denen die eben behandelte Frage nach der Entscheidbarkeit durch endlich viele Operationen nur die letztgenannte war, scheinen mir ein wichtiges, neu zu erschließendes Forschungsfeld zu bilden, und zur Eroberung dieses Feldes müssen wir — das ist meine Überzeugung — den Begriff des spezifisch mathematischen Beweises selbst zum Gegenstand einer Untersuchung machen, gerade wie ja auch der Astronom die Bewegung seines Standortes berücksichtigen, der Physiker sich um die Theorie seines Apparates kümmern muß und der Philosoph die Vernunft selbst kritisiert.

Die Durchführung dieses Programms ist freilich gegenwärtig noch eine ungelöste Aufgabe.

Zum Schlusse möchte ich in einigen Sätzen meine allgemeine Auffassung vom Wesen der axiomatischen Methode zusammenfassen.

Ich glaube: Alles, was Gegenstand des wissenschaftlichen Denkens überhaupt sein kann, verfällt, sobald es zur Bildung einer Theorie reif ist, der axiomatischen Methode und damit mittelbar der Mathematik. Durch Vordringen zu immer tieferliegender Schichten von Axiomen im vorhin dargelegten Sinne gewinnen wir auch in das Wesen des wissenschaftlichen Denkens selbst immer tiefere Einblicke und werden uns der Einheit unseres Wissens immer mehr bewußt. In dem Zeichen der axiomatischen Methode erscheint die Mathematik berufen zu einer führenden Rolle in der Wissenschaft überhaupt.

These specific executions show how various methods of proof are applicable to the same problem; they also suggest to us the necessity to study the essence of mathematical proof itself if such questions as those of decidability by finitely many operations should be answered at all.

All such questions concerning principles as those I have so far characterized—of which only the last mentioned was concerned with the decidability by finitely many operations—seem to me to constitute an important and newly accessible field of research; to capture this field we must—this is my conviction—make the concept of specific mathematical proof itself the object of an investigation, just as the astronomer must take his position into account, the physicist must take care of the theory of his apparatus, and the philosopher criticizes the reason itself.

The implementation of this program is still an unsolved problem at present, of course.

In conclusion, I should like to summarize my general understanding of the axiomatic method in a few lines.

I believe: Everything that can be object of scientific thinking in general, as soon as it is ripe for formation of a theory, runs into the axiomatic method and thereby indirectly to mathematics. Forging ahead towards the ever deeper layers of axioms in the above sense we attain ever deepening insights into the essence of scientific thinking itself, and we become ever more clearly conscious of the unity of our knowledge. In the evidence of the axiomatic method, it seems, mathematics is summoned to play a leading role in science in general.

David Hilbert (1862–1943) studied and started his academic career in Königsberg, the town of his birth. In 1895, he became professor of mathematics in Göttingen, where he stayed for the rest of his life. He was one of the most important mathematicians of the beginning of the 20th century, with seminal contributions to various fields of mathematics. His list of 23 problems, presented at the International Congress of Mathematicians in 1900 in Paris, proved to be highly influential for 20th-century mathematics. One of his many legacies is the promotion of proof theory in the new field of mathematical logic.

Part I
History and Philosophy I(1)

Chapter 2
Hilbert's *Axiomatisches Denken*

Reinhard Kahle and Giovanni Sommaruga

Abstract We discuss some aspects of the impact of Hilbert's talk on *Axiomatic Thinking* in the development of modern mathematical logic.

2.1 Hilbert's 1917 Talk in Context

The work of David Hilbert is often divided in periods where he was concentrating on one topic, not paying much attention to others. The talk on *Axiomatic Thinking* which Hilbert gave on September 11, 1917 at the meeting of the Swiss Mathematical Society marked his "return" to research on foundational questions in mathematics (see [4, p. 200], [7, p. 422f], and [63, p. 150]). This is insofar correct, as it was the moment when Hilbert invited Paul Bernays to return to Göttingen as his collaborator. At that time, he was working out what today we call *Hilbert's Programme*, a research project aiming at providing finitistic consistency proofs for theories formalizing higher mathematics.

Hilbert made his first mark in the foundations of mathematics, when he published, in 1899, his seminal book on *Grundlagen der Geometrie* [28] which gave euclidean geometry a new, modern, and rigorous axiomatization. In the following year, he gave his famous Problems Lecture at the International Congress of Mathematicians in Paris [29]; the first two problems regard foundational questions. In particular

R. Kahle (✉)
Carl Friedrich von Weizsäcker-Zentrum, Universität Tübingen, Keplerstr. 2, 72074 Tübingen, Germany

CMA, FCT, Universidade Nova de Lisboa, 2829-516 Caparica, Portugal
e-mail: reinhard.kahle@uni-tuebingen.de

G. Sommaruga
Eidgenössische Technische Hochschule, Zurich, Switzerland
e-mail: giovanni.sommaruga@phil.gess.ethz.ch

F. Ferreira et al. (eds.), *Axiomatic Thinking I*,
https://doi.org/10.1007/978-3-030-77657-2_2

the second one, the *Consistency of [second-order] Arithmetic*, was instrumental to developing mathematical logic as we know it today; but one also note the sixth problem asking for an *Axiomatization of Physics*. At the subsequent International Congress of Mathematicians, 1904 in Heidelberg, Hilbert gave a first programmatic outline of what consistency proofs could look like [31]. The theory he studied did not contain induction. Poincaré now spotted a vicious circle if one tried to prove the consistency of a theory involving induction by a meta-theoretical argument which would need to make use itself of induction. Apparently, Hilbert was impressed by this objection and stopped, for a moment, this line of research.[1]

Notwithstanding, he pursued his research in logic, notably the study of Frege's work and the *Principia Mathematica* by Whitehead and Russell [64] (see [48, 54]). And Brouwer, according to his own testimony,[2] provided Hilbert with a conceptional solution to Poincaré's vicious circle problem or objection: using *weak* induction in a meta-theory to investigate *strong* induction of an object-theory.

Thus, when he gave his talk in Zurich, Hilbert was sufficiently prepared to seriously restart his foundational research. Already in spring of 1917, on the occasion of a spa stay in Switzerland, he was looking for a possible collaborator and went on a walk with Georg Pólya and Paul Bernays, both working as research assistants in Zurich, in order to discuss his ideas. As Reid [63, p. 150f] reports in her Hilbert biography: "To the surprise of Pólya and Bernays the subject of conversation on the walk to the top of the Zürichberg was not mathematics but philosophy. Neither of them had specialized in that field. Bernays, however, had studied some philosophy and, during his student days at Göttingen, had been close to Leonard Nelson. In fact, his first publication had been in Nelson's philosophical journal. Now, in spite of his quietness, Bernays had much more to say than the usually voluble Pólya." At the meeting in September 1917, Hilbert formally invited Bernays to come to Göttingen, and Bernays accepted.[3]

Hilbert's talk is reprinted in German and in an English translation in this volume [41, Chap. 1]. In the sequel, we will briefly discuss the impact of Hilbert's promotion of *Axiomatic Thinking* in different areas and how this impact is addressed by the contributions in this book.

---

[1] See [7, p. 422]; Hilbert himself commented on this dead end in 1928 [35]: "Eine unglückliche Auffassung POINCARÉs, dieses Meisters mathematischer Erfindungskunst, betreffend den Schluss von n auf $n + 1$, eine Auffassung die überdies bereits zwanzig Jahre früher von DEDEKIND widerlegt war, verrammelte den Weg zum richtigen Vorwärtsschreiten." English translation [39, p. 228]: "An unfortunate view of Poincaré concerning the inference from n to $n + 1$, which had already been refuted by Dedekind through a precise proof two decades earlier, barred the way to progress."

[2] See the footnote to the "first insight" in [9]:"Eine mündliche Erörterung der ersten Einsicht Herrn HILBERT gegenüber hat im Herbst 1909 in mehreren Unterhaltungen stattgefunden." English translation [10, p. 491]: "An oral discussion of the first insight took place in several conversations I had with Hilbert in the autumn of 1909."

[3] Reid, mistakenly, dates this invitation to Hilbert's visit in Zurich in spring.

2.2 Philosophy

At the end of his talk on *Axiomatic Thinking*, Hilbert expressed a bold philosophical thesis [41, p. 19 in this volume]:

> Everything that can be object of scientific thinking in general, as soon as it is ripe for formation of a theory, runs into the axiomatic method and thereby indirectly to mathematics.

To substantiate this claim, there were—at least—two tasks to take care of: one, to give a sound foundation of axiomatic systems, and two, to show how the axiomatic method was to be applied in the sciences.

It is probably fair to say, that later on Hilbert was mainly concerned with the first task which lead in the 1920s to his elaboration of what was to be called *Hilbert's Programme* [68]. The talk itself, however, reserved a considerable part to the second task, that is to the illustration of applications of the axiomatic method first in physics and then also in mathematics. It is the first task, however, which involved specific challenges for philosophy.

2.2.1 *The Role of Logic*

When nowadays we speak of *mathematical logic* we ought to be aware that it is a fairly young concept. Traditionally, logic was regarded as part of philosophy, following Aristotle's fundamental work on syllogisms. Leaving aside the untimely attempts by Leibniz, it was only in the 19th century that Jevons, Boole, and DeMorgan in Britain, Peirce in the US, and Schröder in Germany provided first steps towards linking logic with mathematics. The conceptional breakthrough was due to Frege who succeeded in giving a proper formalization of quantification. However, his formal system for the whole of mathematics, presented in the *Grundlagen der Arithmetik* [19], suffered from the possibility of defining Russell's paradox. At the same time, Cantor's set theory, despite being informal, gave rise to paradoxes impairing the conceptional set-up of set theory.

Hilbert, who was in contact with both, Frege and Cantor, was concerned about these developments, as, on the one hand, he was convinced that Cantor's set theory was an important and fruitful framework for the advance of mathematics—later, in

1925, he spoke emphatically of *Cantor's paradise*—,[4] on the other hand, he saw the security, traditionally associated with mathematical reasoning, in danger.[5]

Thus, he started to investigate the role of logic in mathematics.

Dedekind's work (considering set theory, on principle, as part of logic) and Frege's, as well as Russell and Whitehead's later on, was motivated by *logicism*, i.e., the attempt to reduce mathematics to logic. At first, Hilbert critically and quite positively assessed logicism, but he came, at the latest in 1920, to the conclusion that it was not a viable road to follow.[6]

He recognized though that logic still played a major role as a base of formal frameworks to be used when formalizing mathematics. One should read Hilbert's praise of Frege and Russell in his 1917 talk in this spirit [41, p. 15 in this volume]:

> This road, prepared for a long time — not in the least through the profound investigations by Frege — has finally been pursued by the ingenious mathematician and logician, Russell, with greatest success. In the completion of this extensive enterprise by Russell for the axiomatization of logic one can behold the crowning of the work of axiomatization in general.

When Hilbert followed Peano in adding non-logical components to introduce mathematical concepts, the role of logic proper was restricted to the formalization of the notion of proof.[7] This alone was a revolutionary step. Even if this step was predetermined by Frege and Russell, it only reached maturity in Hilbert's work: by being formalizable, proof itself was turned into a concept accessible to mathematical treatment and investigation. This achievement was expressed by Hilbert in his talk at the International Congress of Mathematicians in 1928 in Rome[8]:

---

[4] See [34]. Hilbert's enthusiasm for Cantor was already expressed in his obituary for his colleague and friend, Minkowski, with whom he had started his studies of set theory in Königsberg [32]: "[Minkowski] war der erste Mathematiker unserer Generation – und ich habe ihn darin nach Kräften unterstützt –, der die hohe Bedeutung der Cantorschen Theorie erkannte und zur Geltung zu bringen suchte." English translation: "Minkowski was the first mathematician of our generation, who realized the high significance of Cantor's theory and who sought to bring it to fruition—and I supported him diligently in this endeavor."

[5] Hilbert expressed this concern in a lecture of 1920 [33, p. 16]: "Insbesondere in der Mathematik, der wir doch die alte gerühmte absolute Sicherheit erhalten wollen, müssen Widersprüche ausgeschlossen werden." English translation: "In particular in mathematics, for which we would like to retain the venerable vaunted absolute certainty, contradictions must be excluded."

[6] Hilbert commented in 1920 [33, p. 33] (published in [40, p. 363]): "Das Ziel, die Mengenlehre und damit die gebräuchlichen Methoden der Analysis auf die Logik zurückzuführen, ist heute nicht erreicht und ist vielleicht überhaupt nicht erreichbar." English translation: "Today, the goal to reduce set theory and thereby the usual methods of analysis to logic is not attained, and it is possibly not attainable at all." In 1928, he expressed it more pithily [36, p. 65]: "Die Mathematik wie jede andere Wissenschaft kann nie durch Logik allein begründet werden." English translation: "Mathematics as any other science can never be founded by logic alone."

[7] See [50]. José Ferreirós, in his contribution to this volume, gives insight in the developments of axiomatics from Dedekind via Peano and Hilbert to Bourbaki [17, Chap. 6].

[8] German original [35]: "Mit dieser Neubegründung der Mathematik, die man füglich als eine Beweistheorie bezeichnen kann, glaube ich die Grundlagenfragen in der Mathematik als solche endgültig aus der Welt zu schaffen, indem ich jede mathematische Aussage zu einer konkret aufweis-

> With this refounding of Mathematics, which can justifiably be called proof theory, I believe to eliminate once and for all the foundational questions in mathematics as such, by transforming every mathematical proposition in a concrete demonstrable and strictly derivable formula and, thus, shifting the entire area to the domain of Pure Mathematics.

And it was later confirmed by Gerhard Gentzen[9]:

> A foremost characteristic of *Hilbert's* point of view seems to me to be the endeavour to withdraw the problem of the foundations of mathematics from *philosophy* and to tackle it as far as in any way possible with methods proper to mathematics.

It is worth discussing to which extent Hilbert succeeded with this aim. But he certainly had a tremendous impact on the development of contemporary *philosophy of mathematics*. This impact was a subject of a panel discussion which took place at the meeting commemorating the centenary of Hilbert's talk in Lisbon on October 11, 2017, with Peter Koellner, Michael Rathjen and Mark van Atten as invited panelists, and which is published in the book [1, Chap. 11].

Also with respect to the role of logic as a formalization device, Hilbert played a decisive role in reforming the shape of logic.[10] It was the Hilbert school which gave predicate logic its finishing touches, the shape that we know today.[11] And it showed how Aristotelian logic could be subsumed under modern mathematical logic, see [42, Chap. 2, Sect. 3]. A specific form of representing Aristotelian relations in the framework of *combinatory logic* [14, 65], an offspring of the Hilbert school's logical investigations, is presented by Erwin Engeler in his contribution to this volume [16, Chap. 5].

And even the implementation of logic as a formal framework underwent an important development in the Hilbert school. The *Hilbert-style calculus*, a term coined by Kleene [51], would have better been named after Frege (who admittedly used an awkward syntax for it). Gentzen [21], a doctoral student of Hilbert (but supervised by Bernays), developed in Göttingen the *sequent calculus* and *natural deductions* as conceptionally fruitful alternatives. In this context, Peter Schroeder-Heister's contribution [67, Chap. 8] links the axiomatic method to recent developments in the area of *proof-theoretic semantics*.

---

baren und streng ableitbaren Formel mache und dadurch den ganzen Fragenkomplex in die Domäne der reinen Mathematik versetze."

[9] English translation from [24, p. 237]; German original [23]: "Ein Hauptmerkmal des *Hilbert*schen Standpunkts scheint mir das Bestreben zu sein, das mathematische Grundlagenproblem der *Philosophie* zu entziehen und es soweit wie irgendmöglich mit den eigenen Hilfsmitteln der Mathematik zu behandeln."

[10] Nelson expressed his astonishment concerning Hilbert's aim to *reform logic* in a letter to Hessenberg, June 1905, cited in [59, p. 166]: "Um den Widerspruch in der Mengenlehre zu beseitigen, will er [Hilbert] (nicht etwa die Mengenlehre sondern) die Logik reformieren." English translation: "To eliminate the contradiction in set theory, [Hilbert] wants to reform (not set theory but) logic."

[11] See the textbook by Hilbert and Ackermann [42]; it is worth mentioning, that the technical elaboration is mainly due to Bernays.

2.2.2 Hilbert's Larger Programme

With Bernays in Göttingen, Hilbert developed in the 1920s his programme to pursue finitistic consistency proofs for formal theories of higher mathematics. This programme relied, of course, on proper formalizations of mathematical areas in the first place. While geometry was the prime example of a formal theory since Euclid's time, arithmetic was surprisingly late subject to formalization in Dedekind's, Peano's, and Frege's work [15, 18, 19, 58]. In contrast to Dedekind and Frege, Peano's work was not based on logicism, but took into account explicitly non-logical components for arithmetic. In the long run, this also was the way pursued by Hilbert. While logic played a role at the base, mathematics proper was given in non-logical terms. And here, axiomatic thinking made its greatest appearance: the non-logical terms received their formal meaning as a *framework of concepts* ("Fachwerk von Begriffen") by non-logical axioms—and by them only.

Hilbert had already masterly carried out this account in his modern axiomatization of euclidean geometry.[12] As geometry can formally be reduced to analysis, he had already in 1900 raised the question of consistency of (second-order) arithmetic.

To prove consistency of an axiomatized theory was one of the two basic requirements given with any such theory. The other was to verify for a set of axioms of an axiomatized theory, whether these axioms were independent from each other or not. Unlike the question of dependence and independence, consistency is of crucial importance as contradictions act like viruses: they infest and imperil the whole body of a theory.[13] In his 1917 talk Hilbert distinguished two approaches to contradictions: first, a "therapeutical approach" (our term), whose purpose it was to eliminate contradictions popping up in a theory, either by axiomatizing the theory (e.g. set theory) or by modifying the theory's axiomatic base. Second, the "preventative approach" (also our term), a more ambitious approach, whose purpose it was to prove that the popping up of contradictions was not possible—that is, the performance of a consistency proof. According to Hilbert, it was 'the principal demand of axiomatics' and the only one ultimately required for mathematics. For many mathematical and for all physical and scientific theories this consistency proof was a relative consistency proof: a proof that a certain theory was consistent provided another theory was (e.g. geometry and analysis). But there were limits to relative consistency proofs. There were no such proofs for the theory of integers and for set theory. The logicist project asked for a "absolute consistency proof" consisting of an axiomatization of logic and a demonstration that the theory of integers as well as set theory were part of logic. We already know what Hilbert's stance to logicism was.

[12] This account profited a lot from previous work by Pasch.

[13] In Hilbert's own words [33, p. 16]: "Ein Widerspruch ist wie ein Bazillus, der alles vergiftet, wie ein Funke im Pulverfass, der alles vernichtet." English translation: "A contradiction is like a bacillus, which poisons everything, as a spark in a powder barrel, which destroys everything." For more on Hilbert's discussion of the paradoxes, see [47].

In his talk of 1917, he also was not merely looking for consistency, but envisaged a much broader potential of the axiomatic method in mathematics, a potential which may be dubbed *Hilbert's Larger Programme*, [41, p. 15 in this volume][14]:

> By closer examination we presently discern that (i) the consistency question of integers and sets is not isolated for itself; rather, it belongs to a great region of the most difficult epistemological questions of specific mathematical coloration. To characterize briefly this region of questions, I mention by name (ii) the problem of principal *solvability of every mathematical question*, (iii) the problem of supplementary *controllability* of the results of a mathematical investigation, (iv) the question of a *criterion for the simplicity* of mathematical proofs, (v) the question of relations between *contentualness* (*Inhaltlichkeit*) and *formalism* in mathematics and logic, and finally (vi) the problem of *decidability* of a mathematical question by a finite number of operations.

The first item (i) was elaborated in the project which is nowadays called *Hilbert's Programme*, and which lead, in particular, to the development of *proof theory* (see the next section). (ii) and (vi) allow to avoid a frequent error concerning Hilbert's conceptional starting point: while (vi) is clearly the decision problem, the missing phrase "by a finite number of operations" in (ii) leaves little doubt that his quest for *solvability of any mathematical question* must not be conflated with the (algorithmic) decision problem, but has to be located on another level and may include meta-theoretical reasoning. (iv) is the only reference in print to *Hilbert's 24th Problem* which he never published, but kept only as an idea in one of his notebooks (see [71] and the collection of papers in [45]). (v) may be considered an anticipation of the distinction of syntax and semantics, in the formal sense developed later by Tarski. Finally, (iii) seems to address concerns with proof-checking.[15]

Thus, when reviewing the execution of Hilbert's Programme in the 1920s and 1930s, one should bear in mind that the axiomatic foundation served much broader purposes than just consistency, many of them followed up with outstanding success. Note also that, according to Hilbert, there was no realization of the *(narrower) Hilbert's Programme* without realization of the *Larger Hilbert's Programme*.

Hilbert was mainly interested in the internal configuration of a framework of concepts. The question, how to justify the specific axioms was not his concern (although he was clearly aware of this question[16]). Thus, Craig Smoryński's contribution, when

[14] The small roman numbers are added by us to simplify referencing.

[15] 50 years later, Bernays wrote an encyclopedia entry for "David Hilbert" [5, p. 500]. Obviously with the 1917 talk at hand, he recalled the given paragraph, however without mentioning (iii). We may take this as an indication that, by 1967, the question of formal proof-checking was considered solved.

[16] See the discussion of *inhaltliche* and *formale Axiomatik* (contentual and formal axiomatics) at the beginning of [43], in particular [43, p. 2]: "Die formale Axiomatik bedarf der inhaltlichen notwendig als ihrer Ergänzung, weil durch diese überhaupt erst die Anleitung zur Auswahl der Formalismen und ferner für eine vorhandene formale Theorie auch erst die Anweisung zu ihrer Anwendung auf ein Gebiet der Tatsächlichkeit gegeben wird." English translation [44, p. 2]: "Formal axiomatics requires contentual axiomatics as a necessary supplement. It is only the latter that provides us with some guidance for the choosing the right formalism, and with some instructions on how to apply a given formal theory to a domain of actuality."

answering the question *Where Do Axioms Come From?* [70, Chap. 10], complements Hilbert's efforts to promote axiomatic thinking. Barry Mazur's contribution [55, Sect. 7.2] contains another complement, namely an overview of the historical development of the concept of axiom.

2.3 Proof Theory

Hilbert's Programme can be divided in two tasks: the first one consists, in accordance with the considerations above, of a technical elaboration of formalization and axiomatization of mathematics.[17] Today, this task is pratically carried out (see the contribution by Wilfried Sieg to this volume [69, Chap. 9]). The second task is to provide consistency proofs for formalized theories. Here, Hilbert made a tactical concession to intuitionism, when he demanded that such consistency proofs should take place in a "meta-mathematics" which had to be intuitionistically acceptable. In fact, Hilbert was—in Weyl's words [75, p. 641]—"more papal than the pope" when he proposed *finitist mathematics* as framework for consistency proofs. What it means to do *meta-mathematics* in an intuitionistic or rather in a constructive setting is spelled out and explained in Lorenz Halbeisen's contribution [27, Chap. 12].

Gödel's incompleteness theorems [26] showed that Hilbert's Programme could not be executed in its original form. It is important to remember though that Gödel could only prove his results after the formal framework had been prepared by research of Hilbert's group in Göttingen; or, as Kreisel [53, p. 177] put it:

> *all* of Gödel's famous work in the thirties had been prepared for him; the problems he solved had been formulated and *made* famous, mainly by Hilbert!

Needless to say, you have to set up the axiomatic method, before you can show its limitations.

Gödel's results were immediately accepted in Göttingen, and Reid reports [63, p. 198]:

> At first [Hilbert] was only angry and frustrated, but then he began to try to deal constructively with the problem. Bernays found himself impressed that even now, at the very end of his career, Hilbert was able to make great changes in his program.

These changes concerned the restriction to finitist mathematics for the meta-theory in which a consistency proof was to be carried out. The restriction was relaxed to constructive mathematics and, based on this change, Gentzen was able to give a constructive consistency proof for first-order Peano Arithmetic in terms of transfinite induction up to ε_0 [22]. There is no doubt, that this proof satisfied Hilbert; but, more importantly, it gave proof theory a new impetus which lead to fruitful developments up

---

[17] It is important to note that Hilbert was only looking for *formalizability* on principle; he never advocated doing mathematics in formal terms; to the contrary, the possibility of formalization, axiomatization together with consistency proofs for the formalized theories would justify the common *informal* reasoning in mathematics.

to the present day. In Part II of this book, modern results in this context are presented in the contributions by Gerhard Jäger [46, Chap. 13] and by Michael Rathjen [62, Chap. 15]; furthermore, Wolfram Pohlers' contribution provides a discussion of the underlying framework for ordinals [61, Chap. 14].

Gödel's results have an immediate corollary, which is underestimated in its historical importance: Second-order logic is not only incomplete, but not even recursively axiomatizable. When Hilbert conceived his programme, he assumed to be able to work in higher-order logic.[18] Due to Gödel's results, we know that first-order logic is the best we can use in axiomatics, when working with finite proofs. Nowadays, it is common sense that mathematics has to be formalized in first-order languages. This limitation of axiomatics is much more profound than the independence of single sentences from specific axiomatic theories—see the discussion of the limits of axiomatics by Bernays, reprinted in this volume [6, Chap. 3].

2.4 Computability Theory

With respect to the decision problem mentioned by Hilbert in his talk,[19] we'd like to put on record that it was Hilbert's tasks leading through formal clarification to the notion of algorithm which we have today.

The *primitive recursive functions* and the Ackermann function exceeding these functions, were defined and studied in the Hilbert school by Bernays, Ackermann, and Péter. General recursive functions were introduced by Herbrand (who closely collaborated with the Hilbert school) and by Gödel, when he was working on his incompleteness results. Turing introduced his concept of Turing machine [72] to (negatively) solve the *Entscheidungsproblem* as presented by Hilbert and Ackermann in [42]. Church worked on his λ-calculus [11, 12] when visiting Göttingen, and, taking up Schönfinkel's [65] pioneering work, Curry even wrote his PhD thesis on combinatory logic [14] under the formal supervision of Hilbert in Göttingen (but, de facto, under Bernays' guidance). Thus, most of the early computability models have their roots in Hilbert's axiomatic enterprise.

Eventually, Alonzo Church proposed the thesis, that the intuitively computable functions were exactly those computable by the equivalent frameworks of λ-calculus, combinatory logic, or Turing machines.[20] Interestingly, this thesis seems to resist a rigid formalization (because of the informal notion of *intuitive computability*); yet, Udi Boker and Nachum Dershowitz' contribution to this volume discusses the possiblities of having a formal proof or disproof of this thesis after all [8, Chap. 20].

Barry Mazur's contribution [55, Sect. 7.4] discusses incompleteness in connection with undecidability, including Hilbert's 10th problem. This not only addresses the

[18] See the discussion in [49].

[19] (vi) in our numbering of the quote above.

[20] Arguably, the model of Turing machines has a distinguished status such that it appears reasonable to speak about the *Church-Turing Thesis* rather than just the *Church Thesis*.

question of decidability—number (vi) in the list above—, but it also shows how aspects of computability are intrinsically connected with the study of axiomatics.

2.5 Axiomatic Thinking at Large

In his 1917 talk, Hilbert appears to distinguish two procedures of the axiomatic method (he seems to use the term 'axiomatic method' for two related, but different procedures) which might be conceived of as two steps of the one axiomatic method. Procedure 1: If one has a framework of concepts, one can form by means of these concepts a few designated propositions from which it is possible to deduce (or, in Hilbert's terms, construct) the entire theory by applying logical principles only. These designated fundamental propositions could be called prima facie axioms. Procedure 2: These prima facie axioms ought to be founded, and they can be "proven" or, as Hilbert put it, reduced to more fundamental propositions of the resp. theory. These more fundamental propositions could be called accordingly secunda facie axioms. This second procedure of the axiomatic method is what Hilbert called the "deeper-laying of foundations" [41, p. 5 in this volume].

Hilbert offered examples of the application of both procedures of the axiomatic method in mathematics, and he also presented a great number of specific examples from physics, reaching even out to economics. Every science is supposed to organize its "facts" in a "framework of concepts" (*Fachwerk von Begriffen*) [41, p. 3 in this volume]:

> If we collate the facts of a specific field of more or less comprehensive knowledge, then we shortly observe that these facts can be set in order. This order occurs invariably with the aid of certain *framework of concepts* such that there exists correspondence between the individual objects in the field of knowledge and a concept of this framework and between those facts within the field of knowledge and a logical relation among concepts. The framework of concepts is nothing but the *theory* of the field of knowledge.
>
> The geometric facts thus order themselves into a geometry, the arithmetic facts into a number theory, the static, mechanic, electrodynamic facts into a theory of statics, mechanics, electrodynamics, or the facts out of physics of gases into a gas theory. The same holds for the fields of knowledge of thermodynamics, of geometric optics, of elementary radiation theory, of heat conduct, or even for probability theory and set theory. Indeed, it holds as well for such specific fields of pure mathematical knowledge as theory of surfaces, Galois theory of equations, theory of primes, and even for none other than some fields of knowledge remotely related to mathematics such as certain segments of psychophysics or economics.

The philosophical rationale of Hilbert's general aim is discussed in Evandro Agazzi's contribution to this volume [2, Chap. 4].

Part III of this book reflects on applications of the axiomatic method within mathematics: John Bell's contribution discusses the method with respect to the notion of *continuity* [3, Chap. 16]; Jean Petitot's contribution uses the Riemann Hypothesis as a case study for axiomatics [60, Chap. 19]; Victor Pambuccian's contribution investigates notions of infinity in an axiomatic perspective [57, Chap. 18]; and Colin

McLarty's contribution considers the mathematical method of *generalization* [56, Chap. 17], likewise prominently advocated by Hilbert.[21]

Hilbert's talk on Axiomatic Thinking gives ample space to physics; and as a matter of fact, he had already prominently placed the *Axiomatization of Physics* as the sixth problem on his list of problems in Paris.[22] Courant [13, p. 162] also reports on the important role played by Hilbert in the development of the structure theory of matter:

> At that time Max Born and Franck had come to Göttingen and Hilbert founded a special seminar with the physicists on the Structure of Matter. The term "structure of matter" really comes from Hilbert's seminar. Many physicists came and all the questions that pertained to the problems of quantum theory and various phases of quantum mechanics and so on were discussed at the seminar. Heisenberg was an important member, Jordan, Born, and Pauli. This was really a very heroic time of modern theoretical physics. The seminar and Hilbert's inspiring interest played a very much greater role in this than the normal art-historical physicist of today knows, or even has the slightest idea about.

In Part IV, Domenico Giulini's contribution [25, Chap. 21] discusses the axiomatic method from the perspective of modern physics.

As an example of a scientific discipline not intimately related to mathematics, Paul Weingartner's contribution [74, Chap. 22] shows how axiomatic thinking can be applied in theology. Applying axiomatic reasoning to theology, as well as to ethics and other scientific disciplines is a main focus of Barry Mazur's contribution [55, Sect. 7.3].

References

1. Panel discussion on the foundations of mathematics. In Fernando Ferreira, Reinhard Kahle, and Giovanni Sommaruga, editors, *Axiomatic Thinking*, volume 1: History and Philosophy, chapter 11. Springer, 2022. This book.
2. Evandro Agazzi. The semantic function of the axiomatic method. In Fernando Ferreira, Reinhard Kahle, and Giovanni Sommaruga, editors, *Axiomatic Thinking*, volume 1: History and Philosophy, chapter 4. Springer, 2022. This book.
3. John L. Bell. Reflections on the axiomatic approach to continuity. In Fernando Ferreira, Reinhard Kahle, and Giovanni Sommaruga, editors, *Axiomatic Thinking*, volume 2: Logic, Mathematics, and other Sciences, chapter 5. Springer, 2022. This book.
4. Paul Bernays. Hilberts Untersuchungen über die Grundlagen der Arithmetik. In *David Hilbert: Gesammelte Abhandlungen*, volume III, pages 196–216. Springer, 1935.

[21] See his Problem Lecture in Paris [29]: "Wenn uns die Beantwortung eines mathematischen Problems nicht gelingen will, so liegt häufig der Grund darin, daß wir noch nicht den allgemeineren Gesichtspunkt erkannt haben, von dem aus das vorgelegte Problem nur als einzelnes Glied einer Kette verwandter Probleme erscheint." English translation [30, p. 443]: "If we do not succeed in solving a mathematical problem, the reason frequently consists in our failure to recognize the more general standpoint from which the problem before us appears only as a single link in a chain of related problems."

[22] We note in passing that he had also subsumed *probability theory* under the sixth problem, a theory which may well serve as prime example for the success of the axiomatic method, if one takes into account its axiomatization by Kolmogorov [52].

5. Paul Bernays. Hilbert, David. In P. Edwards, editor, *Encyclopedia of Philosophy*, volume 3, pages 496–504. Macmillan, 1967.
6. Paul Bernays. Scope and limits of axiomatics. In Fernando Ferreira, Reinhard Kahle, and Giovanni Sommaruga, editors, *Axiomatic Thinking*, volume 1: History and Philosophy, chapter 3. Springer, 2022. This book. Reprinted from Mario Bunge, editor, *Delaware Seminar in the Foundations of Physics*, Springer-Verlag, 1967, pp. 188–191.
7. Otto Blumenthal. Lebensgeschichte. In *David Hilbert: Gesammelte Abhandlungen*, volume III, pages 388–429. Springer, 1935.
8. Udi Boker and Nachum Dershowitz. What is the Church-Turing Thesis? In Fernando Ferreira, Reinhard Kahle, and Giovanni Sommaruga, editors, *Axiomatic Thinking*, volume 2: Logic, Mathematics, and other Sciences, chapter 9. Springer, 2022. This book.
9. Luitzen Egbertus Jan Brouwer. Intuitionistische Betrachtungen über den Formalismus. *Koninklijke Akademie van wetenschappen te Amsterdam, Proceedings of the section of sciences*, 31:374–379, 1927. Translation in part in [10].
10. Luitzen Egbertus Jan Brouwer. Intuitionistic reflections on formalism. In J. van Heijenoort, editor, *From Frege to Gödel*, pages 490–492. Harvard University Press, 1967. English translation of §1 of [9].
11. Alonzo Church. A set of postulates for the foundations of logic. *Annals of Mathematics (2)*, 33(2):346–366, 1932.
12. Alonzo Church. A set of postulates for the foundations of logic (second paper). *Annals of Mathematics (2)*, 34(4):839–864, 1933.
13. Richard Courant. Reminiscences from Hilbert's Göttingen. *The Mathematical Intelligencer*, 3(4):154–164, 1981. Edited transcript of a talk given at Yale University on January 13, 1964.
14. Haskell B. Curry. Grundlagen der kombinatorischen Logik. *Amer. J. Math.*, 52:509–536, 789–834, 1930.
15. Richard Dedekind. *Was sind und was sollen die Zahlen?* Vieweg, 1888.
16. Erwin Engeler. Aristotle's relations: An interpretation in combinatory logic. In Fernando Ferreira, Reinhard Kahle, and Giovanni Sommaruga, editors, *Axiomatic Thinking*, volume 1: History and Philosophy, chapter 5. Springer, 2022. This book.
17. José Ferreirós. The two sides of modern axiomatics: Dedekind and Peano, Hilbert and Bourbaki. In Fernando Ferreira, Reinhard Kahle, and Giovanni Sommaruga, editors, *Axiomatic Thinking*, volume 1: History and Philosophy, chapter 6. Springer, 2022. This book.
18. Gottlob Frege. *Die Grundlagen der Arithmetik*. Wilhelm Koebner, 1884. Reprinted, M. & H. Barcus, Breslau, 1934 and Olms, Hildesheim, 1961. English translation in [20].
19. Gottlob Frege. *Grundgesetze der Arithmetik. Begriffsschriftlich abgeleitet*, volume I–II. Hermann Pohle, Jena, 1893/1903.
20. Gottlob Frege. *The Foundations fo Arithmetic*. Blackwell, 1950. Translation of [18] by J. L. Austin.
21. Gerhard Gentzen. Untersuchungen über das logische Schließen I, II. *Mathematische Zeitschrift*, 39:176–210, 405–431, 1935.
22. Gerhard Gentzen. Die Widerspruchsfreiheit der reinen Zahlentheorie. *Mathematische Annalen*, 112:493–565, 1936.
23. Gerhard Gentzen. Die gegenwärtige Lage in der mathematischen Grundlagenforschung. *Forschungen zur Logik und zur Grundlegung der exakten Wissenschaften, Neue Folge*, 4:5–18, 1938. also in *Deutsche Mathematik*, 3:255–268, 1939. English translation [24].
24. Gerhard Gentzen. The present state of research into the foundations of mathematics. In M. E. Szabo, editor, *The Collected Works of Gerhard Gentzen*, Studies in Logic and the Foundations of Mathematics, pages 234–251. North-Holland, 1969. English translation of [23].
25. Domenico Giulini. Axiomatic thinking in physics – essence or useless ornament? –. In Fernando Ferreira, Reinhard Kahle, and Giovanni Sommaruga, editors, *Axiomatic Thinking*, volume 2: Logic, Mathematics, and other Sciences, chapter 10. Springer, 2022. This book.
26. Kurt Gödel. Über formal unentscheidbare Sätze der Principia Mathematica und verwandter Systeme. *Monatshefte für Mathematik und Physik*, 38:173–198, 1931.

27. Lorenz Halbeisen. A framework for metamathematics. In Fernando Ferreira, Reinhard Kahle, and Giovanni Sommaruga, editors, *Axiomatic Thinking*, volume 2: Logic, Mathematics, and other Sciences, chapter 1. Springer, 2022. This book.
28. David Hilbert. Grundlagen der Geometrie. In *Festschrift zur Feier der Enthüllung des Gauss-Weber-Denkmals in Göttingen, herausgegeben vom Fest-Comitee*, pages 1–92. Teubner, 1899.
29. David Hilbert. Mathematische Probleme. Vortrag, gehalten auf dem internationalen Mathematiker-Kongreß zu Paris 1900. *Nachrichten von der königl. Gesellschaft der Wissenschaften zu Göttingen. Mathematisch-physikalische Klasse aus dem Jahre 1900*, pages 253–297, 1900. Reprinted in [37, pp. 290–329]; English translation in [30].
30. David Hilbert. Mathematical Problems. Lecture Delivered Before the International Congress of Mathematicians at Paris in 1900. *Bulletin of the American Mathematical Society*, 8:437–479, 1902. English translation of [29].
31. David Hilbert. Über die Grundlagen der Logik und der Arithmetik. In Adolf Krazer, editor, *Verhandlungen des Dritten Internationalen Mathematiker-Kongresses in Heidelberg vom 8. bis 13. August 1904*, pages 174–185. Leipzig, 1905.
32. David Hilbert. Hermann minkowski. *Göttinger Nachrichten, Geschäftliche Mitteilungen*, pages 72–101, 1909. Reprinted in: *Mathematische Annalen*, 68:445–471, 1910 and in [37, pp. 339–364].
33. David Hilbert. Probleme der mathematischen Logik. Vorlesung Sommersemester 1920, Ausarbeitung von N. Schönfinkel und P. Bernays (Bibliothek des Mathematischen Instituts der Universität Göttingen), 1920. Published in [40,342–371].
34. David Hilbert. Über das Unendliche. *Mathematische Annalen*, 95:161–190, 1926. English translation in [38].
35. David Hilbert. Probleme der Grundlegung der Mathematik. In *Atti del Congresso Internazionale dei Matematici (Bologna, 3–19 settembre 1928)*. Nicola Zanichelli, 1928a. Offprint; English translation in [39].
36. David Hilbert. Die Grundlagen der Mathematik. *Abhandlungen aus dem mathematischen Seminar der Hamburgischen Universität*, 6(1/2):65–85, 1928b.
37. David Hilbert. *Gesammelte Abhandlungen*, volume III: Analysis, Grundlagen der Mathematik, Physik, Verschiedenes, Lebensgeschichte. Springer, 1935. 2nd edition 1970.
38. David Hilbert. On the infinite. In Jean van Heijenoort, editor, *From Frege to Gödel*, pages 367–392. Harvard University Press, 1967. English translation of [34].
39. David Hilbert. Problems of the Grounding of Mathematics. In P. Mancosu, editor, *From Brouwer to Hilbert*, pages 227–233. Oxford University Press, 1998. English translation of [35].
40. David Hilbert. *Lectures on the Foundations of Arithmetic and Logic 1917–1933*, volume 3 of *David Hilbert's Lectures on the Foundations of Mathematics and Physics, 1891–1933*. Springer, Edited by W. Ewald and W. Sieg edition, 2013.
41. David Hilbert. Axiomatisches Denken. In Fernando Ferreira, Reinhard Kahle, and Giovanni Sommaruga, editors, *Axiomatic Thinking*, volume 1: History and Philosophy, chapter 1. Springer, 2022. This book. German original with English translation; the German text is reprinted from *Mathematische Annalen*, 78:405-415, 1918; the English translation of Joong Fang was first published in Joong Fang, editor, *HILBERT–Towards a Philosophy of Modern Mathematics II*, Paideia Press, Hauppauge, N.Y. 1970.
42. David Hilbert and Wilhelm Ackermann. *Grundzüge der theoretischen Logik*, volume XXVII of *Die Grundlehren der mathematischen Wissenschaften in Einzeldarstellungen*. Springer, 1928.
43. David Hilbert and Paul Bernays. *Grundlagen der Mathematik I*, volume 40 of *Die Grundlehren der mathematischen Wissenschaften in Einzeldarstellungen*. Springer, 1934. 2nd edition 1968.
44. David Hilbert and Paul Bernays. *Grundlagen der Mathematik I/Foundations of Mathematics I*. College Publications, 2011. Bilingual edition of Prefaces and §§1–2 of [43].
45. Inês Hipolito and Reinhard Kahle. Theme issue on "The notion of simple proof – Hilbert's 24th problem". *Philosophical Transactions of the Royal Society A*, 377, 2019.
46. Gerhard Jäger. Simplified cut elimination for Kripke-Platek set theory. In Fernando Ferreira, Reinhard Kahle, and Giovanni Sommaruga, editors, *Axiomatic Thinking*, volume 2: Logic, Mathematics, and other Sciences, chapter 2. Springer, 2022. This book.

47. Reinhard Kahle. *David Hilbert über Paradoxien*, volume 06-17 of *Pré-publicações do Departamento de Matemática, Universidade de Coimbra*. 2006.
48. Reinhard Kahle. David Hilbert and the Principia Mathematica. In N. Griffin and B. Linsky, editors, *The Palgrave Centenary Companion to Principia Mathematica*, pages 21–34. Palgrave Macmillan, 2013.
49. Reinhard Kahle. Is there a "Hilbert Thesis"? *Studia Logica*, 107(1):145–165, 2019. Special Issue on General Proof Theory. Thomas Piecha and Peter Schroeder-Heister (Guest Editors).
50. Reinhard Kahle. Dedekinds Sätze und Peanos Axiome. *Philosophia Scientiæ*, 25(1):69–93, 2021. Special Issue on Giuseppe Peano and his School: logic, epistemology and didactics. Paola Cantù and Erika Luciano (Guest Editors).
51. Stephen Cole Kleene. *Introduction to Metamathematics*. D. Van Nostrand Company, New York, 1952.
52. Andrei Kolmogorow. *Grundbegriffe der Wahrscheinlichkeitstheorie*. Springer, 1933.
53. Georg Kreisel. Gödel's excursions into intuitionistic logic. In Paul Weingartner and Leopold Schmetterer, editors, *Gödel Remembered*, volume IV of *History of Logic*, pages 65–186. Bibliopolis, 1987.
54. Paolo Mancosu. The Russellian influence on Hilbert and his school. *Synthese*, 137:59–101, 2003.
55. Barry Mazur. Notes for a seminar in axiomatic reasoning. In Fernando Ferreira, Reinhard Kahle, and Giovanni Sommaruga, editors, *Axiomatic Thinking*, volume 1: History and Philosophy, chapter 7. Springer, 2022. This book.
56. Colin McLarty. Abstract generality, simplicity, forgetting, and discovery. In Fernando Ferreira, Reinhard Kahle, and Giovanni Sommaruga, editors, *Axiomatic Thinking*, volume 2: Logic, Mathematics, and other Sciences, chapter 6. Springer, 2022. This book.
57. Victor Pambuccian. Varieties of infiniteness in the existence of infinitely many primes. In Fernando Ferreira, Reinhard Kahle, and Giovanni Sommaruga, editors, *Axiomatic Thinking*, volume 2: Logic, Mathematics, and other Sciences, chapter 7. Springer, 2022. This book.
58. Ioseph Peano. *Arithmetices Principia Nova Methodo Exposita*. Bocca, Augustae Taurinorum, 1889.
59. Volker Peckhaus. *Hilbertprogramm und Kritische Philosophie*, volume 7 of *Studien zur Wissenschafts-, Sozial- und Bildungsgeschichte der Mathematik*. Vandenhoeck & Ruprecht, Göttingen, 1990.
60. Jean Petitot. Axiomatics as a functional strategy for complex proofs: the case of Riemann Hypothesis. In Fernando Ferreira, Reinhard Kahle, and Giovanni Sommaruga, editors, *Axiomatic Thinking*, volume 2: Logic, Mathematics, and other Sciences, chapter 8. Springer, 2022. This book.
61. Wolfram Pohlers. On the performance of axiom systems. In Fernando Ferreira, Reinhard Kahle, and Giovanni Sommaruga, editors, *Axiomatic Thinking*, volume 2: Logic, Mathematics, and other Sciences, chapter 3. Springer, 2022. This book.
62. Michael Rathjen. Well-ordering principles in proof theory and reverse mathematics. In Fernando Ferreira, Reinhard Kahle, and Giovanni Sommaruga, editors, *Axiomatic Thinking*, volume 2: Logic, Mathematics, and other Sciences, chapter 4. Springer, 2022. This book.
63. Constance Reid. *Hilbert*. Springer, 1970.
64. Bertrand Russell and Alfred North Whitehead. *Principia Mathematica*, volume 1. Cambridge University Press, 1910. vol. 2, 1912; vol. 3, 1913.
65. Moses Schönfinkel. Über die Bausteine der mathematischen Logik. *Mathematische Annalen*, 92:305–316, 1924. English translation [66].
66. Moses Schönfinkel. On the building blocks of mathematical logic. In Jean van Heijenoort, editor, *From Frege to Gödel: A Source Book in Mathematical Logic, 1879–1931*. 73, pages 355–366. Harvard University Press, 1967. English translation of [65].
67. Peter Schroeder-Heister. Axiomatic thinking, identity of proofs and the quest for an intensional proof-theoretic semantics. In Fernando Ferreira, Reinhard Kahle, and Giovanni Sommaruga, editors, *Axiomatic Thinking*, volume 1: History and Philosophy, chapter 8. Springer, 2022. This book.

68. Wilfried Sieg. Hilbert's programs: 1917-1922. *The Bulletin of Symbolic Logic*, 5(1):1–44, 1999.
69. Wilfried Sieg. Proofs as objects. In Fernando Ferreira, Reinhard Kahle, and Giovanni Sommaruga, editors, *Axiomatic Thinking*, volume 1: History and Philosophy, chapter 6. Springer, 2022. This book.
70. Craig Smoryński. Where do axioms come from? In Fernando Ferreira, Reinhard Kahle, and Giovanni Sommaruga, editors, *Axiomatic Thinking*, volume 1: History and Philosophy, chapter 10. Springer, 2022. This book.
71. Rüdiger Thiele. Hilbert's twenty-fourth problem. *American Mathematical Monthly*, 110(1):1–24, 2003.
72. Alan M. Turing. On computable numbers, with an application to the Entscheidungsproblem. *Proceedings of the London Mathematical Society*, 42:230–265, 1936.
73. Jean van Heijenoort. *From Frege to Gödel*. Harvard University Press, 1967.
74. Paul Weingartner. Axiomatic thinking – applied to religion. In Fernando Ferreira, Reinhard Kahle, and Giovanni Sommaruga, editors, *Axiomatic Thinking*, volume 2: Logic, Mathematics, and other Sciences, chapter 11. Springer, 2022. This book.
75. Hermann Weyl. David Hilbert and his mathematical work. *Bulletin of the American Mathematical Society*, 50(9):612–654, 1944.

Reinhard Kahle is Carl Friedrich von Weizsäcker Professor for Philosophy and History of Science at the University of Tübingen. Before he hold professorships in Mathematics at the University of Coimbra and at the Universidade Nova in Lisbon, at the end as full professor for Mathematical Logic. He is fellow of the *Académie Internationale de Philosophie des Sciences*. His main research interests are proof theory and the history and philosophy of modern mathematical logic, in particular, of the Hilbert School. He has (co-)edited more than ten books and special issues as, for instance, Gentzen's Centenary: The quest for consistency (Springer, 2015) and The Legacy of Kurt Schütte (Springer, 2020), both togehter with Michael Rathjen.

Giovanni Sommaruga did his studies in philosophy and philosophical and mathematical logic at the University of Freiburg (Switzerland), Stanford University and the University of Siena. In 1996 he became assistant professor in logic and philosophy of science at the Albert Ludwig University Freiburg (Germany), and since 2008 he has been senior scientist in philosophy of the formal sciences (logic, mathematics, theoretical computer science) at ETH Zurich. His main research interests are in philosophy and foundations of mathematics, in the history of mathematical logic, and more recently in the history and philosophical issues concerning computability and information in theoretical computer science.

Chapter 3
Scope and Limits of Axiomatics

Paul Bernays

Abstract Reprinted with permission from *Delaware Seminar in the Foundations of Physics* edited by M. Bunge, Springer-Verlag, 1967, pp. 188–191.

When today one speaks of axiomatics to a public familiar with mathematics, there often seems to be not so much a need for recommending axiomatics as to warn against an overestimation of it.

In fact, there are today mathematicians for whom science begins only with axiomatics, and there are also mathematically minded philosophers, especially in CARNAP's school, who regard axiomatization as belonging to the construction of scientific languages. Thus science is then regarded as being essentially deductive, whereas in fact only some sciences, and even so in an advanced stage, proceed mainly by derivation.

There are indeed empirical sciences where certain kinds of objects, for instance plants, are investigated as to their various forms and behaviors, as to the ways of their occurring as well as to the structure of their development. Here one obviously has first to care for acquiring a sufficient empirical material, and a premature axiomatization might divert one from a sufficient exploration of what is to be found in nature, or at least it may restrict this exploration.

On the other hand we have to be conscious that collecting and presenting material has a scientific value only if this material gives rise to suitable conceptions, classifications, generalizations, and assumptions. A mere material, without any theoretical aspect joined to it, cannot even be kept in mind, at least if it has a considerable extent. It is true that tables of collected material can prove to be highly valuable even in a way not originally taken into account; but certainly this takes place only if the material is brought in connection with directive ideas, i.e. with some conceptuality.

Now, whenever some conceptuality develops, when the question is one of putting results together, or of evaluating heuristically a problem situation, or also of subjecting assumptions to empirical tests, or even of investigating their inner consistency—in all such cases axiomatic thinking can be very fruitful. In fact it is then a matter of making clear what exactly the results are yielding, or what properly the problem

P. Bernays (1888–1977)
Eidgenössische Technische Hochschule, Zürich, Switzerland

F. Ferreira et al. (eds.), *Axiomatic Thinking I*,
https://doi.org/10.1007/978-3-030-77657-2_3

situation is, or in what the assumptions strictly consist. Such a strengthened consciousness is valuable whenever the danger exists that we may be deceived by vague terminology, by ambiguous expressions, by premature rationalizations, or by taking views for granted which in fact include assumptions. Thus, the distinction between inertial mass and gravitational mass makes it clear that their equality is a physical law—something which might be overlooked by speaking of mass as the quantity of matter. Various instances of such a kind were especially considered by ERNST MACH.

Another sort of occurrence of axiomatics in the development of science is the presentation of a new directive idea in axiomatic form. A famous instance of this is the presentation of thermodynamics by CLAUSIUS, starting from the two main laws; and, of course, the classical instance of NEWTON's presentation of his mechanics cannot be omitted in this connection.

For both the said methods of applying axiomatics one can find also examples in biology, in particular the theory of descent and genetics, as well as in theoretical economics.

All the axiomatic systems considered so far have the common trait that they sharpen the statement of a body of assumptions, either for a whole theory or for a problem situation, and of being embedded in the conceptuality of the theory in question; eventually, then, this conceptuality might be somewhat sharpened by it. We might perhaps call them "material" or "pertinent" axiomatics.

In connection with pertinent axiomatics it may be mentioned that some important enterprises in theoretical science, in particular in physics, have an axiomatic aspect—namely all those where some discipline comes to be incorporated in another discipline, or where two disciplines are contracted into one theory. Well known instances are the incorporation of thermodynamics into mechanics by the methods of probability theory and statistics, and the embedding of optics in the theory of electromagnetic waves, as well as the joining of physical geometry with gravitation theory in general relativity theory. Of course we do not have here simply axiomatics, since the said reductions were partly motivated and also had to be justified by experimental results.

Nevertheless such reductions can also be considered under the aspect of that kind of axiomatic investigation which HILBERT called "the deepening of the foundations" of a theory and of which he gave many instances in his article "Axiomatisches Denken".

Examples of such a deepening are: the proof of the solvability of every algebraic equation, with one unknown, in the domain of complex numbers—though the modern algebraists do not admit this theorem as an algebraic one, at least in its usual presentation—, and the reduction of the theory of real numbers to that of natural numbers with the addition of the concept of number set or else that of a denumerable sequence.

The methods used in the deepening of foundations gave rise to a kind of axiomatics different from pertinent axiomatics and which one might call definitory or descriptive axiomatics, to which belong abstract disciplines developing the consequences of a structural concept described by axioms. Examples of such concepts are those of group, lattice, and field.

There is also the possibility of combining such concepts. Thus the metric continuum can be characterized as an ordered field whose order satisfies the Dedekind condition of absence of gaps.

A remarkable circumstance in descriptive axiomatics is that there is a multiplicity of equivalent characterizations. Thus lattice theory can be axiomatized just as well with the concept of equality or with the relation "sub" as a fundamental predicate. In group theory we can postulate the existence of a right-side unit element and right-side inverse; or else we may postulate the solvability of each of the equations $ax = b$ and $ya = b$. Hence axiomatization contains elements of arbitrariness. The newer axiom systems for geometry, in particular those of HILBERT and of VEBLEN, are also descriptive, in contrast with the axiom system of PASCH, which is intended as a pertinent one. In the geometrical axiom systems a great freedom in the choice of kinds of individuals and the fundamental relations is to be found.

You may wonder why I have not yet mentioned the most famous instance of axiomatics: Euclid's *Elements*. But from the point of view of method this instance is not simple.

According to the traditional view the Euclidean system is to be regarded as a pertinent axiomatics, yet not starting from empirical facts but rather from intuitively evident facts. You know that there has been and there still is much debating about geometrical self-evidence. Yet however this question might be decided, Euclid's enterprise cannot be understood as attempting to secure a high degree of intuitive evidence.

The way of taking the geometric axioms as purely hypothetical, as it si predominant in to-day's mathematic, is, at least, strongly prepared in Euclid's Elements. Only the axioms on quantities (the κοιναὶ ἔνοιαι) have here a separate status (It is taken as self-evident, and it is not even axiomatically formulated, that the lengths of segments, the angles, areas and volumes are quantities satisfying those axioms).

It was also a merit of Euclid's axiomatization that attention was called to the special role of the parallel axiom and thereby mathematicians became later on induced to discover non-Euclidean geometries.

In contemporary mathematics geometry is treated in many different disciplines: analytic geometry—in particular vector geometry—, differential geometry, projective geometry, algebraic geometry, and topology. FELIX KLEIN distinguished the kinds of geometry by the group of transformations which perserve the properties studied by the discipline.

Under the influence of symbolic logic one has passed from axiomatization to formalization. In the latter,

(1) the possible forms of statements in a theory are delimited apriori, and
(2) the logical inferences are subjected to explicit rules.

It is possible to formalize in this way the existing proofs in number theory, infinitesimal analysis, set theory, geometry, etc.

Nevertheless formalized axiomatics cannot fully replace descriptive axiomatics. The reason is that the sharpening of the concepts of predicate, function, sequence, and set, which is brought about by formalization, includes a restriction of these concepts to the effect that interpretations of the formal systems are possible which are excluded by the corresponding original descriptive axiomatics.

This deficiency consists for the formalization of any axiom system wherein one of the said concepts is used, as in the induction axiom of number theory or in the continuity axiom of analysis.

It appears in a twofold way:

(a) in a "syntactic" way, by the impossibility of formally proving certain number-theoretic theorems, which hold true upon the consistency of the formal axiom system,—as it was stated by GÖDEL.
(b) in a "semantic" way, by the existence of "non-standard models", i.e. models differing from the structure to be described by the axiom system,—as stated first by SKOLEM.

So there seems to be a limit of strict axiomatization in the sense that we either have to admit a certain degree of imprecision, or to be unable adequately to characterize what we mean, for instance, by well ordering, continuity, and number series.

Paul Bernays (1888–1977) was Professor of Higher Mathematics at ETH Zurich. After his studies in mathematics and philosophy at the University of Göttingen, he was *Privatdozent* at the University of Zurich when, in 1917, he joined David Hilbert, again in Göttingen, to work on the Foundations of Mathematics. This collaboration culminated in the publication of the two volumes *Grundlagen der Mathematik* in 1934 and 1939. Dismissed from the University in Göttingen in 1933 because of his Jewish origin, he returned to Zurich and started lecturing at ETH. His research extended from proof theory to set theory and the philosophy of logic and mathematics.

Chapter 4
The Semantic Function of the Axiomatic Method

Evandro Agazzi

Abstract Since ancient Greek mathematics the axiomatic method had essentially a *deductive function* in the sense of granting the *truth* of the propositions of a certain discipline by starting from primitive propositions that were "true in themselves" and submitting them to a truth-preserving formal manipulation. A new perspective emerged in the foundational research of Peano's school and Hilbert's view of mathematical theories. According to this view, the whole axiomatic system provides a sort of global meaning that offers the possibility of understanding he sense of a theory. and also to find possible models for it. We can call this the *semantic function* of the axiomatic method because it offers the two basic dimensions of semantics, i.e. *sense and reference* (according to the Fregean perspective). Intellectual intuition is no longer needed for providing the objects of mathematical theories and the requirement of truth is replaced by the requirement of *consistency*. This approach has immediate consequence of an ontological nature, i.e. regarding the *kind of existence* of mathematical entities (Is consistency a sufficient condition for existence in mathematics?).

4.1 The Axiomatic Method as Paradigm of Perfect Science

The axiomatic method is almost as old as the very concept of genuine knowledge in Western culture. Indeed, when Plato posed in the dialogue *Menon* the question of how to distinguish opinion (*doxa*) from knowledge, that he calls "science" (*episteme*) he affirmed that truth is not sufficient, since there are also true opinions, and maintained that, in the case of science, truth must be accompanied by a "discourse giving its reasons".[1] In what such a discourse should consist he tried to explain in the *Teetetus* and *Protagoras* and this task was fully accomplished in Aristotle's *Organon*,

[1] Plato, *Menon*, 98a.

E. Agazzi (✉)
Center for Bioethics of the Panamerican University of Mexico City, Mexico, Universities of Genoa, Genoa, Italy
e-mail: evandro.agazzi@gmail.com

F. Ferreira et al. (eds.), *Axiomatic Thinking I*,
https://doi.org/10.1007/978-3-030-77657-2_4

whose *Prior Analytics* present the doctrine of formal deduction (syllogistics), and the *Posterior Analytics* propose the model of a genuine science, in which the formal deduction is applied to self-evident *first principles.* This amounted to proposing the axiomatic method as the paradigm of any genuine science, and we are entitled to call this method "axiomatic" because Aristotle himself mentions the axioms, as they were understood by the mathematicians, in a passage of his *Metaphysics* where he assigns to the philosopher the task of finding the genuine first principles of any science, from which also the axioms of the mathematicians should have their foundation:

> We must state whether it belongs to one or to different sciences to inquire into the truths which are in mathematics called axioms, and into substance. Evidently, the inquiry into these also belongs to one science, and that is the science of the philosopher; for these truths hold good for everything that is, and not for some special genus apart from others……And for this reason no one who is conducting a special inquiry tries to say anything about their truth or falsity,-neither the geometer nor the arithmetician…. Evidently then it belongs to the philosopher, i.e. to him who is studying the nature of all substance, to inquire also into the principles of syllogism…. This is the philosopher, and the most certain principle of all is that regarding which it is impossible to be mistaken; for such a principle must be both the best known (for all men may be mistaken about things which they do not know), and non-hypothetical. For a principle which every one must have who understands anything that is, is not a hypothesis; and that which every one must know who knows anything, he must already have when he comes to a special study. Evidently then such a principle is the most certain of all; which principle this is, let us proceed to say. It is, that the same attribute cannot at the same time belong and not belong to the same subject and in the same respect; we must presuppose, to guard against dialectical objections, any further qualifications which might be added. This, then, is the most certain of all principles, since it answers to the definition given above. For it is impossible for any one to believe the same thing to be and not to be, as some think Heraclitus says. For what a man says, he does not necessarily believe; and if it is impossible that contrary attributes should belong at the same time to the same subject (the usual qualifications must be presupposed in this premise too), and if an opinion which contradicts another is contrary to it, obviously it is impossible for the same man at the same time to believe the same thing to be and not to be; for if a man were mistaken on this point he would have contrary opinions at the same time. It is for this reason that all who are carrying out a demonstration reduce it to this as an ultimate belief; for this is naturally the starting-point even for all the other axioms.[2]

This text is particularly interesting not only because it gives to non-contradiction the rank of *first principle* of every human knowledge, but also because its characteristic of being indisputable and absolutely true is presented as the characteristic that should possess also the axioms of the special sciences, each being concerned only with a particular *genus* of things, but all having inherited from the first principle the privilege of absolute certainty and truth. This view is so fascinating that it has remained the ideal of perfect knowledge for the rest of Western philosophy (for example, it was explicitly advocated in Fichte's short work *On the concept of the doctrine of science or the so-called philosophy* of 1794).[3] This ideal model constitutes the *analytic* conception of knowledge that envisages it as a solid building in which no obscurity exists, since truth is secured by demonstration and meaning

[2] Aristotle, *Metaphysics* IV, 3. Trans by D.Ross.

[3] This work is included in Fichte [12], pp. 29–sgg.

through definition. Demonstration and definition, however, are particular intellectual processes that have the common feature of 'constructions' proceedings from some starting points.

4.2 Mathematics as Prototype of an Axiomatized Science

Mathematics has been considered the most perfect science through almost the entire history of Western civilization, mainly for its being able to establish absolutely true propositions, and to grant this truth by means of indisputable proofs, precisely in keeping with the ideal model outlined by Aristotle. In fact, the truth of a mathematical proposition was meant to be granted either by absolute intellectual evidence, or by a stringent logical argument making this truth follow necessarily from the truth of certain evident propositions. In this sense, proof or 'demonstration' has been classically understood as 'reduction to evidence', and the idea of a demonstrative science was presented along the lines of Aristotle's *Posterior Analytics*, in particular, in Euclid's *Elements*, whose structure constituted the model of what is known as the axiomatic method. In its 'classical' form this method can be sketched as follows: All sound knowledge must be formulated in unambiguously meaningful and reliably true propositions; unambiguous meaning can be established through clear *definitions*, and reliable truth through correct *proofs*. But not everything can be defined, nor can everything be proved, and 'starting points' are needed in both cases. These starting points are of two sorts: primitive notions—whose meaning is evident, i.e. immediately known—which are needed for the construction of the definitions (which then 'transfer' the meaning to further notions); and primitive propositions (also called axioms or postulates depending on the case)—whose truth is evident, i.e. immediately known—for the construction of the proofs (which then 'transfer' the truth to further propositions). The immediateness of the meaning and of the truth of the primitive elements was strictly related to the uniqueness of the *referents* or objects about which the solid knowledge was intended to be.

This method, that has never been rejected in the whole history of Western mathematics, has two sides: the strictly methodological one—which we can qualify as "syntactic"—regards the way of constructing mathematical theories, and presenting the results of mathematical investigation, independently of the content or meaning of the particular domain at issue (and, therefore, independently also of the truth or falsity of the sentences); the second one—that we can qualify as "semantic"—concerns the meaning and truth of the propositions of the discourse and, therefore, necessarily entails the consideration of the 'content' of the discourse itself. The interplay between these two sides is very significant, and was noted already in Antiquity. For example, one of the few successful efforts for proving the 5th Euclidean postulate on the parallels (proposed by Posidonius and Geminus in the I century B.C.) consisted in changing the definition of parallel lines by stating that two parallel lines are 'equidistant', in the sense that the points of the one keep the same distance from the points of the other or, in other words, that the parallel to a given straight line r

is the geometric locus of the points having the same distance from r. It was noted, however by Proclus (II century A.D.) that this definition implicitly presupposed that this locus is still a straight line, and this is actually a new postulate introduced in elementary geometry, which is not more evident than the 5th postulate itself. In other words, changing the meaning of a concept (semantic level) can entail changing the structure of the accepted sentences (syntactic level). During the nineteenth century several such interplays occurred, especially regarding the concept of the infinite. For example, the introduction of the *concept* of limit (as the value that a function or sequence 'approaches' as the input or index approaches some value) was clearly introduced by Cauchy in his *Cours d' Analyse* of 1821[4] and became the cornerstone for the rigorous construction of calculus (and mathematical analysis in general), being used to *define* continuity, derivatives, and integrals and allowing for the *proof* of numberless theorems. Even more instructive can be the 'paradox' of infinite sets, for which the 'whole' can be proved 'equivalent' with some of its 'parts', something contrary to spontaneous intellectual intuition and that was explicitly rejected in the last of Euclid's axioms, but was recognized as a puzzling property of natural numbers by Galileo in the seventeenth century. When the *concepts* of proper subset of a set, and of equivalence among sets were codified as a bijective correspondence between the elements of two sets, the fact that a set can be equivalent with some of its proper subsets, instead of being rejected, was promoted to the role of *definition* for the concept of infinite set and opened the way to the *definition* of the notion of cardinality and the construction of transfinite set theoretic arithmetic, with all its impressive *theorems*. The admission of this definition, however, was the consequence of admitting in mathematics the legitimacy of conceiving of *actual infinity*, something that the elaboration of the concept of limit had carefully avoided (remaining within the framework of *potential infinity*), and this was a genuine *philosophical* step in the sense of the Aristotelian text quoted above, a step that was crucial in the (not uncontroversial) acceptance of Cantor's set theory.

The global sense of the foregoing remarks amounts to giving to the semantic level a priority with respect to the construction of an axiomatic system, that can take place when its primitive concepts have been duly singled out and elaborated, so that the syntactic structure of the system does not affect its semantic 'stable' ground and precondition. This approach reflects a well-known standard partition of logic, which starts with a doctrine of concepts, then goes on with a doctrine of judgments (that are conceived as a combination of concepts which receives from them its meaning) and ends up with a doctrine of argumentation (that consists in a derivation of judgments that should preserve the truth of the premises down to the conclusion).

[4] Cauchy [9].

4.3 Non-contradiction Replaces Intuitive Evidence

The adoption of this ideal portrayal was challenged by the unsuccessful efforts of *directly* proving the Euclidean 5th postulate (whose immediate evidence was doubtful) exclusively from the other four postulates, without the help of additional explicit or implicit postulates. The turn occurred with the *Euclides vindicatus* of Girolamo Saccheru [31],[5] who tried to prove this postulate *indirectly*, that is, by showing that the negation of the postulate entailed a *logical* contradiction (independently of the fact that certain anti-intuitive theorems were deduced during this poof). This pioneering approach became customary about one century later, when the first systems of non-Euclidean geometries were explicitly constructed by Bolyai and Lobachevsky and their legitimacy was confined to their non-contradiction (or consistency as we call it today), whereas their repugnancy with ordinary geometrical intuition was no longer considered as a valid objection. Geometric intuition, as we have seen, was traditionally considered to be the means for affirming the *evidence* of the geometrical propositions, and hence also the warranty of their truth. Therefore, giving up the requirement of geometrical intuition amounted to giving up the requirement of truth for the geometrical propositions and (since truth was understood as a correspondence of the meaning of the propositions with the real structure of the domain of the geometrical entities or objects) giving up the requirement of intuition also entailed that geometrical propositions are neither true nor false and that geometry does not have a proper domain of objects that it investigates. This situation—that we have roughly described as the outcome of a historical process that was indeed very complex and rich of important conceptual and mathematical innovations—used to be expressed by saying that geometry in particular, but also mathematics in general, consisted of several formal systems *devoid of meaning*. This phrase, of course, cannot be taken literally, because any discourse that is really devoid of meaning is simply not understandable, whereas the formal systems of mathematics, in spite of being sometimes abstract, are understandable. Therefore, this phrase must be understood in the sense that they have not just one fixed *domain of referents*, as it was believed in the tradition (so that, for instance, geometry studied the properties of figures in space, arithmetic the properties of natural numbers, etc.). Moreover, this alleged absence of meaning was quickly considered as the great advantage of being open to different *interpretations*, and this amounted to the recognition of a considerable fruitfulness of this formal outlook. We intend now to consider what types of meaning can be attributed to the axiomatic formal systems.

4.4 The Syntactic Meaning

It is patent, from the historical sketch offered above, that the axiomatic method has had traditionally the function of 'organizing' the content of a given discipline or

[5] Saccheru [31].

science, by imposing on it a *structural ordering* or 'systematization' that should be totally explicit and transparent. This amounted, first, to clearly specifying the class of the *terms* (in which a subclass of 'primitive' terms and one of 'derived' terms had to be distinguished) and indicating the procedure for such a derivation (that was qualified as *definition*). The second step consisted in specifying the class of the *sentences* (in which a subclass of 'primitive' sentences and one of 'derived' sentences have to be distinguished) and indicating the procedure for such a derivation (that was qualified as *proof*). The primitive sentences were called *axioms* or *postulates*, and the derived sentences were called *theorems*. Regarding 'proof', it was tacitly understood that it should consist in the adoption of the deductive rules of *formal logic*.

This general scheme was supposed to be fully understandable and, therefore, meaningful, and even the 'boxes' indicated with the labels "definition" and "proof" were supposed to be 'open' and available as a kind of toolboxes from which certain adequate instruments could be safely taken for the construction of the axiomatic system. The reason for which we have called "syntactic" this meaning is that it does not consider the 'content' of the particular discourse to which it can be applied. We might have called it also a *functional* meaning, because what one needs to understand is only the function attributed to the primitive or derived terms or sentences, what is the function of the different elements constituting the procedures of definition and of formal logical deduction, And all these functions can be (and have actually been) described in detailed and 'meaningful' ways in the treatises of ancient and modern logic. In particular, at the end of the nineteenth century they had been expressed in accurate *symbolic form* that assured their total explicitness and made patent their functional nature.

The spontaneous question is whether this syntactic or functional meaning exhausts the whole of the meaning that is appropriate to attribute to mathematics, and the crisis of the mathematical intuition that we have described certainly encouraged the view that there is nothing more than a syntactic meaning in mathematics. In fact, when we say that we *understand* a certain sentence, we normally mean that we can form a *representation* of what it says, and this representation amounts concretely to having a certain *intuition* of what the sentence is supposed to express. Hence, if we are in a situation in which such intuitions are *not* (or no longer) available, we are compelled to admit that—in the said situation—the sentence is *devoid of meaning*, and this was precisely the historical situation regarding mathematics at the end of the nineteenth century. Therefore, we can take seriously the famous statement of Russell when he defined mathematics as "The subject in which we never know what we are talking about, nor whether what we are saying is true."[6]Nevertheless, this conclusion was not that pessimistic, and the same Russell offered a construction of the meaning and content of mathematics as founded on pure logic already with *The Principles of Mathematics (*1903)[7] and especially, in collaboration with Whitehead, with *Principia Mathematica* (1910–1913).[8] As is well-known, this was that reduction

[6] Russell [29].

[7] Russell [30].

[8] Whitehead and Russell [34].

of mathematics to logic which is known as "logicism" and whose chief representatives were Frege, Russell and Withehead, who considered logic itself as the formal expression of the objective contents of thought. For this reason the syntactic meaning was considered sufficient in order to attribute to mathematics also a kind of objective content, that was the way of functioning of the human thinking.

This, however, was not the only viable solution for giving to the axiomatic method the capability of serving a better understanding and construction of genuinely mathematical theories. Giuseppe Peano and his school, in the last decade of the 19th and the first decade of the twentieth century, had created logical and symbolic tools for a critical analysis and rigorization of the standard branches of mathematics, and used the axiomatization as a means for showing how the contents of such theories could be preserved and developed by means of different axiomatizations, that is, by using different primitive concepts and axioms (for example, taking as primitive notions for elementary geometry those of point and segment, or of point and motion).[9] What is interesting in this position is the application of formal logical tools to the study of 'concrete' mathematical disciplines, but at the same time remaining 'agnostic' as far as the 'ontology' of such theories is concerned and, as a consequence, also 'independent' of any particular ontology. In this approach, one single theory is susceptible of several axiomatizations, whereas the formalistic view held the inverse approach: one single axiomatic system can receive several interpretations. Due to this difference, we can understand why Peano's school produced the most satisfactory clarification of the function of the axiomatic method in the determination of the *meaning* of mathematical concepts and statements, but was not interested in the determination of the *referents* of mathematical theories. It was also, not concerned with the problem of consistency of the axiom systems, a problem which—on the contrary—became crucial for the formalistic school and even moved it to attribute to formal systems also a *referential* power, as we shall later consider more closely.

4.5 The Semantic Meaning (or Sense)

The crisis of mathematical intuition was, in a more precise sense, the admission of the incapability of mathematics to capture its intended *referents* because those intuitions were considered as representations or images of the said referents (such

[9] Already in 1889 Peano had inaugurated a critical examination of basic mathematical theories, such as arithmetic and elementary geometry, by means of rigorous formal and symbolic axiomatizations (see Peano [25, 26]). Also the idea that axiom systems constitute a contextual definition of their own primitive notions originated in his school, and was developed particularly by Alessandro Padoa (see [23]) and Mario Pieri (see [27, 28]). However, since these views were also accepted by Hilbert, and applied by him in his famous book *The Foundations of Geometry* (see [19]), they became much better known as expressing the Hilbertian and in general the formalistic conception of mathematics. However, it must be said that Peano and his school never advocated a truly 'formalistic' view of mathematics, formal tools being for them essentially instruments for the analysis and exposition of mathematical theories, and not for creating them.

as figures in space, natural numbers, rational and real numbers, functions, and so on) and this is the sense of Russell's statement quoted above. As a consequence, efforts aiming at giving back to mathematics or, more precisely, to an axiomatic system, a meaning should consist in showing that it can have some referents. This was actually Hilbert's original position that considered this attribution of meaning as an *interpretation* of the axioms on certain domains of objects, indicating a way that was later developed in mathematical logic with the Tarskian creation of model theory. There is, however, also a different way of conceiving meaning that was commonly recognized in medieval logic and was recovered in the Fregean distinction between *Sinn* and *Bedeutung*,[10] two terms that are usually translated today as *sense* and *reference*. The sense is the cognitive content of an act of thinking (a thought), and the referent is the object to which this thought is applied. In ordinary language, the notion of "meaning" is ambiguous, being understood sometimes as sense, and sometimes as reference, whereas it is correct to say that both sense and reference are distinct parts of meaning, which we can call the "semantic meaning" and the "referential meaning" respectively.[11] According to this distinction, we can recognize that Peano's school has given a fundamental contribution in clarifying a significant performance of the axiomatic method in relation with the semantic meaning, whereas Hilbert's school has rather conceived this performance in the sense of the referential meaning.

What do they mean "point", "line", "surface", "angle", "circle" and possibly other primitive nations of elementary geometry? Euclid had offered for them commonsensical characterizations that could not be considered definitions (since these notions are primitive), but simply 'clarifications' obtained by embedding them in a broader linguistic context. The original idea developed within Peano's school was that this meaning should result from the web of mutual relations that connect these notions in the different axioms, so that the whole axiom system provides a sort of definition of these notions. The novelty of this proposal can be appreciated if we consider that in traditional logic only one type of definition was canonically recognized, that is, the definition *per genus et differentiam* that consists, in practice, in the inclusion of a concept in a broader one. This happened because definitions were meant to provide the meaning of new concepts by 'tracing' them back to the meaning of already known primitive concepts. Moreover, this 'tracing back' had been conceived, especially since the seventeenth century, in a rather special sense, i.e. as a decomposition or analysis of 'complex' concepts into their 'simple' or not further analyzable 'elements'. However, it was by no means granted that such primitive elements need be equipped with a really clear meaning. Just to give a few examples, concepts which have been accepted as primitive include 'thought', 'motion', 'light' (Locke), 'time' (Pascal), 'object' and 'concept' (Frege), which are far from being unproblematic. The way to overcome this difficulty—proposed by Peano's school and applied with great precision and strength by Hilbert—was to reverse this strategy: the meaning of the primitive concepts is to be conceived as globally given by the whole context of the

[10] See Frege [13].

[11] See for details Agazzi [3], essentially taken up in Agazzi [4], Chap. 4.

formal axiomatic system in which they occur. This is the core idea of an 'axiomatic definition', which expresses an authentic *semantic function* of the axiomatic method.

This new way of conceiving the axiomatic method was prepared by Moritz Pasch,[12] and completely developed by Peano and his school between 1889 and 1899 (i.e., in the decade preceding the first publication of Hilbert's *Foundations of Geometry*). In particular Mario Pieri offered the first explicit presentation of this new approach in a paper with the significant title *On elementary geometry as a hypothetical-deductive system: a monograph of the point and motion* (1899),[13] It is worth noting that in this paper Pieri coined the phrase "hypothetical-deductive system" for characterizing mathematical theories that became later very common in the literature. Pieri also presented his view in French at the first International Congress of Philosophy held in Paris in 1900.[14] He explicitly defended the idea that the meaning of the primitive concepts is 'defined' through the postulates. This "definition through postulates," was often called later "implicit definition" but this latter expression, was not considered a very satisfactory way of characterizing this contextual interdependence of meanings, and nowadays it is used in a much more restricted and technical sense in mathematical logic. The expression "structural definition" has also been used more recently.[15]

The first full-fledged treatise which adopts this new view was David Hilbert's famous book *Foundations of Geometry* of 1899[16] that differed from the traditional geometry textbooks of the time, not because it advocated a new geometry (in fact its content is still comparable with that of ordinary geometry), but for its conceiving of the axiomatic method in a new manner. "Point," "straight line," "plane," and so on were no longer presented as names for specific geometrical entities, but as terms the meanings of which were 'contextually' defined by all the axioms, and which were thus capable of having as referents whatever objects that could satisfy those same axioms. The great impact of this treatise (in which, in particular, the mutual independence of the different axioms was accompanied by the outline of alternative geometries that could be constructed) and the great authority of Hilbert as mathematician, had the effect that this semantic function of the axiomatic method is often presented as Hilbert's original view.[17]

---

[12] Pasch [24].

[13] Pieri [27].

[14] Pieri [28].

[15] See, for instance, Sieg [33].

[16] Hilbert [19].

[17] In fact, Peano's booklet of 1889, *I principi di geometria logicamente esposti* presents the undeniable novelty of adopting a complete symbolization and even of including logic along with the specific geometric constituents of his formalization (something that Hilbert did not do in his *Foundations of Geometry*). Nevertheless one must admit that the reading of his pages of exclusively symbolic proofs constitutes perhaps a difficulty, rather than an advantage, for appreciating the semantic potentialities of his approach. Moreover, the limited extension of this work (40 pages) compared with the rich display of original vistas contained in Hilbert's treatise, and also the fact of being written in Italian, explains why it did not receive the attention it deserves (at variance with

But was such a move really correct? Criticisms, as we shall see, were formulated against it, but we must understand the reasons for its having become widely accepted. It was the natural consequence of the new way of conceiving the axiomatic method in mathematics, when this method was no longer regarded simply as a tool for introducing deductive order in a discipline, but as something that was able to determine, at least to some extent, the very objects of the discipline. The difference between these two positions is rather patent. If one considers axiomatization as a way of deductively ordering a discipline, one regards at least a certain number of the terms occurring in the axioms as *names* for the entities which the discipline is supposed to describe, and the 'meaning' of these terms may be regarded as their reference to these objects. But if one considers axiomatization as something that must 'create' a certain discipline, no objects are presupposed as existing, and the axioms must in a way be able to have meaning even without having, properly speaking, a reference.

If this is the case, the meaning must necessarily arise from the reciprocal links that the different concepts have with each other; and, if a question of reference is advanced, it can only concern the possibility of discovering some *structure* of objects the relations of which can be put into correspondence with the *links* between the concepts expressed in the axioms, so as to be faithfully represented by them.

A serious criticism to the formalistic approach was expressed by Frege: in several papers devoted to Hilbert's *Foundations of Geometry* which appeared in the *Jahresbericht der deutschen Math.-Vereinigung* in the years 1903 and 1906, as well as in a couple of letters to Hilbert.[18] Frege correctly pointed out that the totality of the postulates may at most define a 'second order' concept (of which the primitive concepts occurring in the postulates are so to speak the ingredients), but cannot establish the meaning of these concepts themselves.[19] Frege's criticism remained uninfluential (owing to the growing favor of the formalist trend in mathematics), and at most led later to a 'readjustment' of the issue. As was suggested by Bernays in a review of the then newly discovered correspondence between Frege and Hilbert (published in the *Journal of Symbolic Logic* 7, 1942, pp. 92–93), the postulates of elementary

the international recognition of Peano's results in analysis published in the *Mathematische Annalen* or with his celebrated axiomatization of elementary arithmetic).

[18] See Frege [14].

[19] To be a little more precise: Frege pointed out that, in order to apply Hibert's proposal rigorously, one had to consider the axioms as open sentences containing unquantified predicate variables without any precise meaning. To consider the whole axiom system as a definition of these predicates amounts to introducing a second-order predicate, which is satisfied 'by definition' if and only if the different axioms are simultaneously satisfied on a given domain when they are replaced by suitable predicate constants. Therefore what Hilbert actually does is to confuse the definition of one n-place second-order predicate with an allegedly reciprocal definition of n first-order predicates. This criticism of Frege was overlooked, and even considered as evidence of his being unable to enter the new spirit of mathematics. Indeed this evaluation is to be found even in the words of such an attentive and competent scholar and logician as Heinrich Scholz, who wrote: "nobody nowadays doubts that Frege, who had himself created so many novelties in the field of the classical concept of science, was no longer able to understand the radical Hilbertian restructuration of this concept of science, so that his critical remarks, though being in themselves highly acute and still worth being read, must be considered as essentially pointless" (Scholz [32], p. 222).

geometry, for example, represent an explicit definition not of the single concepts occurring in them, but of the concept of Euclidean three-dimensional space.

Even so, Frege's correct criticism has not been met; and if we pay it the attention it deserves, we must at least refrain from saying that the axiomatic context (or any linguistic context) *entirely* determines the meaning of the concepts. We are certainly entitled to say that this meaning depends *also* on the context on an *intensional* level; but this dependence cannot mean the dissolution of the meaning in the context, otherwise no meaning at all could emerge. This, by the way, is why the thesis of total 'meaning variance' so widely advocated in philosophy of science is already untenable for semantic reasons.

We are not interested here in discussing the issue of the semantic 'stability' which must exist to some extent even if context is given its fullest role. What we may say here is that every concept enters into a scientific theory equipped with a meaning whose structure is well articulated and depends on many factors. It is therefore wrong to say that terms receive their meanings totally and only through the theoretical context. This is actually an authentic formalistic fallacy which goes back to Carnap's proposal to consider physical theories as interpreted logical formal calculi (see Carnap [8]). In fact empirical theories do not begin to exist as formal systems, but may at most be 'formalized' after they have attained a certain stage of development. At this stage it may also be possible to detect the 'variance' of meaning which occurs as a consequence of a term's being located in a different context, but this 'variance' is always partial. Therefore, when a term is used in a particular sentence, it is normally used only according to a part of its meaning, and it may well happen that the part concerned is not affected by 'meaning variance.' This general remark applies also to non-empirical theories such as those of mathematics, and was not overlooked by Peano and his followers, who proposed different axiomatizations of really existing mathematical theories, respecting a certain fundamental meaning for primitive concepts that received additional 'variable' parts of meaning when placed within the various axiomatic contexts.

4.6 The Referential Meaning

The last considerations remind us that the axiomatic method has never been considered in the whole history of Western civilization as something like a 'language game', but always as a means for securing stability and certainty to human knowledge whenever it concerns certain domains of objects. This is what Hilbert and Bernays have called sometimes the "existential aspect" of axiomatizations, which we could also consider as a 'descriptive' aim of axiomatization, as distinct (but not separated) from its 'constructive' aim. The persistence of this existential dimension in spite of all formalistic developments is particularly evident in Hilbert's paper *Axiomatisches Denken*[20] that corresponds to the content of a lecture given by him in Zurich at a

[20] Hilbert [20].

meeting of the Swiss Mathematical Society in 1917. In this paper Hilbert advocates a universal adoption of the axiomatic method in the field of all exact sciences, offering examples taken from physics and mathematics and showing, in such a way, that also mathematical theories were considered, by him, as discourses having their own specific domain of objects or *referents*, as we would say according to the Fregean distinction mentioned above. This fact is particularly significant if we consider that it comes 17 years after the publication of the *Foundations of Geometry* that seemed to inaugurate a trend of dissolution of the idea of the 'existence' of mathematical entities, whereas the subjacent problem was that of the 'kind of existence' or of the 'ontological status' of the mathematical entities.

This problem has boiled down that 'quest for rigor' that characterized mathematics during the whole nineteenth century and which had found in the 'arithmetization of analysis' one of its most salient expressions. This arithmetization consisted in considering certain *numbers* (concretely speaking, the real numbers and, by extension, also the complex numbers) as the proper 'objects' of analysis. The 'real' status of the real numbers, however, was far from clear, and well-known proposals were advanced around 1872 for 'constructing' the real numbers starting from the rational numbers and, going back in this genealogy, starting from the natural numbers. At this point, however, the question regarding the nature of the natural numbers could no longer be escaped, and this is particularly clear in the reflection of Richard Dedekind who had proposed in 1972 his construction of the real numbers as "cuts" in the domain of the rational numbers,[21] but felt obliged to enter the challenging question of the nature of the natural numbers 16 years later.[22] In that second paper he essentially proposed a characterization of the natural numbers as constructions of pure logic, being in such a way a forerunner of the 'logicist' foundational school, whose different approaches were formulated by Frege[23] and Russell.[24] Nevertheless, he wanted explicitly to prove the existence of an infinite set, through a very peculiar proof that, though continuing to consider mathematical entities as contents of thought, recognized in the process of indefinite reflections of thoughts one on the other, the 'real' example of an infinite set. Dedekind's logical construction of the natural numbers, however, could not avoid serious perplexities, the most relevant of which were expressed by Leopold Kronecker who advocated for the natural numbers an 'existence' independent of logic, whose role is essential, however, for the genetic construction of the other types of numbers starting from that 'natural' basis.[25]

The admission of a basic structure of referents was also a condition that Hilbert himself has used for proving the consistency of axiomatizations and, in particular,

---

[21] Dedsekind [10].

[22] Dedekind [11].

[23] Frege [15].

[24] Russell [30].

[25] This is the sense of Kronecker's famous statement, "God has created the natural numbers; all the rest is man's work." It is a rhetorical expression of his conviction that the whole of mathematics must be reducible to arithmetic. These views are clearly presented e.g. in Kronecker [22] and were the reason for his strongly opposing Cantor's set theory.

was applied in his *Foundations of Geometry* whenever he wanted to prove the consistency of the different axiom systems obtained in the course of his critical discussions: on those occasions he provided an interpretation of the geometric axioms on some special structure of the real numbers and could say that this was an adequate warranty of consistency in virtue of the consistency of real numbers analysis that, in turn, was secured by the 'genetic' construction starting from the natural numbers that was considered a pacific result. But was elementary arithmetic consistent? This question was considered by him of a primary importance, and actually appears as the second (immediately following the issue of the continuum hypothesis) in the famous list of "mathematical problems" proposed by him at the International Congress of Mathematics celebrated in Paris in 1900. Therefore, it is not accidental that the problem of finding a 'direct' proof of the consistency of elementary arithmetic reappears as the fundamental aim of "Hilbert's Program" elaborated in the 1920s.[26]

That Program was, in a way, the culmination of the formalistic view of mathematics, since the whole mathematical domain was considered as a web of formalized axiomatizations (including arithmetic) in which room was given to a great variety of mathematical concepts, including "Cantor's paradise" of the infinite, and logical consistency was the only warranty required for the acceptance of such theories, A warranty that was looked for by investigating not the *contents*, but the *proofs* used in mathematics (*Beweistheorie*). Such an investigation could not dispense with having its own referents, and they were the most rudimentary manipulations of a few material *signs* according to a finite number of instructions. This "finitary" procedure was hoped to be sufficient for *showing* that no formal contradiction could follow from the axioms of arithmetic but, as is well-known, Gödel's results[27] proved that those methods (being formalizable within arithmetic itself) could not produce the hoped result. From that moment on, it has become customary to offer as a warranty for the consistency of a system of axioms an interpretation that shows that they have a model on some admitted structure of referents, in the sense that they are *satisfied* in that structure according to a certain interpretation.

This last remark invites us to a critical reflection on a judgment that was rather common in the writings of the early logical empiricists, when they affirmed the superiority of the 'modern' way of conceiving mathematics in comparison with the naïve traditional one, by stressing that the ancients were totally unaware of the need of proving the internal consistency of their axiom systems. We must simply note that the ancients considered the axioms and postulates of a mathematical theory as

[26] It is worth noting that the problem of the consistency of axiom systems was explicitly studied already in Peano's school, though with no such an urgency that it received later within Hilbert's formalistic approach.In March 1903 Padoa published a long deep article in *L'Enseignement Mathématique* on compatibility of axioms and non-contradictory axioms. He wrote the article in response to the problems that had arisen in set theory and became clear especially in the following work of Russell. This led to a lively debate in which also Hilbert participated. Padoa published a 106-page book *La logique déductive dans sa dernière phase de développement* in 1912; this book consisted in the reprint of three separate articles published in *Revue de metaphysique et de morale* in 1911 and 1912, and a foreword to the book was written by Peano.

[27] See Gödel [17].

true propositions concerning the specific objects of that theory. A correct logical deduction from true premises always leads to true consequences. Therefore, since a contradictory sentence is always false, no contradictions could derive from a system of true axioms. If the ancients had posed the problem of a consistency proof of the axioms of mathematics they would have been really naïve. This reasoning is very simple, but not simple minded, and reflects that relation of language, thought and ontology that was summarized in the Aristotelian text which we quoted at the beginning, and has received extensive treatment in the history of philosophy. Therefore, it is no wonder that also today the common strategy for securing the consistency of an abstract axiom system is that of finding for it a model in a domain of 'given' referents.

This is by no means an accidental or contingent situation but is required by the very notion of interpretation, that is, of the notion that has allegedly eliminated the old fashioned view that mathematical theories have an ontological domain of reference. Indeed, in order to give whatever interpretation, it is not sufficient to have a system of symbols to be interpreted, but it is indispensable to explicitly indicate a *non empty set of entities* on which the symbols have to be interpreted. This, by the way, is the philosophically interesting core of Hilbert's turn to the investigation of *proofs* that hopefully required only the sensory perception of material symbols and their material manipulations, something that (as he repeatedly stressed) had nothing to do with logic. The hope was that one could *show* that even a rudimentary string of symbols expressing a contradiction could not be obtained by using those finitary procedures.[28]

4.7 Mathematical Existence and Consistency

The fact that the warranty of consistency for a formal system is usually looked for by finding a suitable interpretation on a structure of 'given' referents does not have, in itself, a great ontological significance, simply because the ontological status of such referents is dependent on the particular condition in which they are considered as 'given'. The less objectionable position is that of considering such interpretations as proofs of *relative consistency*, in the sense that the consistency of the formal system under consideration is shown to be entailed by the consistency of the *theory* of the structure in which the model is found. This, by the way, was the idea that inspired the construction of the Euclidean models of non-Euclidean geometries in the nineteenth century, when the initial search for contradictions in those geometries

[28] We cannot enter the complex philosophical issues involved in the different approaches briefly mentioned here, regarding in particular the 'contentual ' view of mathematical theories and axiomatizations, different from the 'existential axiomatics' proper and from 'formal axiomatics'. These philosophical issues were considered at the beginning of Hilbert-Bernays [21] and were discussed with great accuracy and depth in several papers by Bernays, of which it may be sufficient to mention Bernays [6, 7]. I have devoted myself a historical–critical analysis of the development of the axiomatic method in Agazzi [1].

appeared fruitless.[29] The conclusion that these scholars attained was not that the different non-Euclidean geometries were consistent, bust simply that *if* the traditional Euclidean geometry is consistent, *then* also the non-Euclidean are consistent. In the same vein one should say that the fact of finding a model for a formal system by suitable arrangements of real numbers or, in the last analysis, in the domain of natural numbers actually amounts to 'discharging' the consistency proof on the consistency proof for elementary arithmetic, and this, as we have seen, was clear to Hilbert already in 1900 and he did not look for a more fundamental domain of ontologically given referents (not even in logic) and invented twenty years later the new strategy of proof theory.

A challenging question now surfaces: can the axiomatic method, beside the important semantic function that it performs in the analysis of *sense*, also perform a referential function, that is, 'produce' a structure of *referents*? Of course, not unconditionally, but only at the condition of being consistent? Or, to put it in a sharper form "Does any consistent set of axioms have a model"? The answer to this question is not straightforward because of the vagueness of the very notion of model or, more precisely,of the indeterminacy of the 'ontological status' of the model. For example, Henri Poincaré, who has often stressed that logic and intuition have a different function in mathematics (because logic presides over demonstration, and intuition is necessary for invention), and has also proposed one of the most elegant and simple Euclidean models for non-Euclidean hyperbolic geometry, maintained on other occasions that every consistent set of axioms has a model.[30]

An interesting affirmative answer to the above question was offered by Leon Henkin in a celebrated paper of 1949[31]were he proved that an arbitrary set of sentences constructed using first-order logic and supposed consistent, has a model. From this result easily follows the semantic completeness of first order logic, which is just a special case of the general fact that if whatever consistent set S of sentences formulated by means of a certain logical calculus has a model, that calculus is semantically complete (that is, it allows to formally deduce all the logical consequences of whatever set of sentences). The curious fact, however, is that the (skillful) strategy adopted for constructing the said model consisted in taking as universe U of the objects on which the sentences of the consistent set S are interpreted, the set T of the terms of the language itself, and stating that any term is interpreted on itself. Then a sentence containing the terms $t_{1....}t_e$ is declared satisfied in U if and only if it belongs to S.[32]

[29] The first proposal of such a model was made by Eugenio Beltrami [5], and was followed by other similar models proposed by Cayley, Poincaré, Klein.

[30] We do not enter here into details since the position of Poincaré on this issue, which was presented especially in *Science and Method* and in several articles published in *Revue de métaphysique et de moral* was rather complex, and complicated also by the fact of being in part motivated by a dispute with scholars like Couturat, Le Roy, Hilbert, as well as by an unfriendly attitude towards formal symbolism in logic.

[31] Henkin [18].

[32] To be slightly more precise: the supposed consistent set S is embedded in a succession of larger and larger sets obtained by adding new sentences and new individual constants in ways that always

The traditional view that consistency is a necessary (and tacitly presupposed) condition for the mathematical truth can be easily translated in the claim that every set S of propositions that admits a model is consistent: this is an easy well-known theorem of mathematical logic. The most challenging enterprise is to prove the reciprocal thesis, that is, that consistency is also a sufficient condition for mathematical truth. Such a claim could sound rather innocent and almost a way for hinting at the "abstract" nature of mathematics, but it takes up a more engaging sense if we translate it in the claim that every consistent set of sentences possesses a model. In fact, this formulation endows consistency with something like an "ontological creativity" that is far from self-evident. There is no more the possibility of escaping the difficulty by saying that mathematical truth is "abstract", because it should be in any case a truth concerning a certain structure of "objects" (possibly of abstract objects, but always of objects). How can consistency be endowed with an ontological creativity? How can a system of arbitrary, though consistent, conditions ensure that there exists in the world a structure of objects about which it turns out to be true?[33]

Henkin's brilliant proof does not answer these questions. Indeed one cannot overlook that in the semantics of mathematical logic and model theory, as they have been started by Tarski and then remained standard, the domain of objects on which the interpretation of a formal language is made was considered as a structure *different from the language* that is going to be interpreted. This corresponds to the intention (explicitly expressed by Tarski) of respecting the fundamental characteristics of the notion of truth, which is the property of a discourse saying something about its referents and not about itself. The self-referential features of the interpretation of S, on the contrary, fails to respect this requirement since the domain on which S is interpreted does not possess any structure of its own, but is structured according to the prescriptions contained in S itself. Therefore, the situation obtained could be described, perhaps a little ironically, as follows. Our theorem proves that any consistent set of propositions of first order logic describes a "possible world". We ask "Which world"? and the answer is "Of course, the world described by these

---

preserve consistency. One obtains in such a way a set S* which is consistent and enjoys several properties, among which the fact that if it has a model, this will also automatically be a model of S. Then a model is found for S^*, that is also a model of S The standard procedures for obtaining a model are: (i) that a "domain of objects" ω be provided; (ii) that an interpretation j of the language L of S* on ω be defined such that; (iii) a model of S* can result, i.e. that the sentences of S* are satisfied on ω. This should rigorously express, at least according to Tarski's original proposals, the idea that the formal expressions of S* became true about ω under this interpretation. This domain is for Henkin, in the simplest case, the set V of the individual variables of L (or the set T of the equivalence classes of the closed terms of L in the case of a first order logic with identity), and the subsequent steps of the interpretation are articulated in such a way that functions and relations are declared to hold of their arguments if and only if the corresponding sentence belongs to S*.

[33] This distinction was also central to Kant's philosophy and backed his distinction between thinking and knowing: not whatever that can be "thought" can also be "known", or, to put it differently, "thoughts" are not "objects" and since a claim of existence usually regards the domain of objects, it follows that simple thoughts are not sufficient for stating the existence of anything. Even in the most favorable situations (such as that of consistency) simple thought is not endowed with ontological creativity.

propositions"! In more serious terms we must recognize that even in the case of first order logic the consistency of a set of propositions cannot secure the existence of a domain of objects ,ontologically given ' in an autonomous way, about which these propositions tell the truth. Hence, even in this privileged case.

We shall conclude with a more general remark. As we have already noted, it holds in general that, if any consistent set of sentences expressed by means of a formal calculus L has a model, that calculus is semantically complete. Hence if it were true that every consistent set of mathematical propositions possesses a model we should conclude that all logical calculi are semantically complete, but we know that this is the case only for the first order predicate calculi (including sentential calculi). Therefore, we should conclude that the ontological creativity of consistency can at best be envisaged in the case of systems of propositions formulated in first order languages, whereas it could not hold for systems of propositions for whose formulation a more powerful language is needed.

These remarks are sufficient for our problem: we were trying to see whether the *existence* of mathematical entities could be considered superfluous and replaced by the simple *consistency* of the systems of mathematical propositions, though accepting the traditional notion of truth that entails reference to ,objects'. This could be possible if consistency were able, either to ,generate 'by itself certain objects or to offer a warranty concerning the existence of objects capable of 'making true' the consistent propositions. We see now that this condition could obtain, at best, for particular systems of propositions, depending on the language used, and this leaves us puzzled, for one does not see how the existence of the mathematical objects could depend on the structure and richness of the language that speaks about them. But we have seen that, even in the privileged case in which mathematical logic seems to indicate that consistency is endowed with an ontological power—that is, the case of first order logic, for which it has been proved that every consistent set of sentences has a model—no genuine ontological creativity is present, since the model is artificially created within the language itself.

References

1. Agazzi, E. (1961), *Introduzione ai problem dell'assiomatica*, Vita e Pensiero, Milano
2. Agazzi, E. (2011), Consistency, Truth and Ontology, *Studia Logica* 97/1, pp. 7–29
3. Agazzi, E. (2012), Meaning between sense and reference:Impacts of semiotics on philosophy of science. *Semiotica.* Special issue *"Semiotics and Logic"*, 188–1/4 (2012), pp. 29–50
4. Agazzi, E. (2014), *Scientific Objectivity and its Contexts*, Springer, Cham, Heidelberg, New York, Dordrecht, London
5. Beltrami, E. (1868), Saggio di interpretazione della geometria non euclidea, *Giornale di matematiche.* VI, pp. 284–322.
6. Bernays, P. (1922), Ueber Hilberts Gedanken zur Grundlegung der Arithmetik, JDMV 31,1922, pp. 10–19. (Lecture delivered at the Mathematikertagung in Jena, September, 1921).
7. Bernays, P. (1950), Mathematische Existenz und Widerspruchsfreiheit, in *Etudes de philosophie des sciences en hommage à Ferdinand Gonseth,* Editions du Griffon, Neuchatel, pp. 11–25

8. Carnap, R. (1934), *Die logische Syntax der Sprache*, Springer, Wien. Engl. trans. Kegan Paul, London, 1937
9. Cauchy, A.L. (1821), *Coursa d'analyse de l'École Royale Polytechnique,* Denur, Paris.. Repr. by Wissenschaftliche Buchgesellschaft, Darmstadt, 1968.
10. Dedekind, R. (1872), *Stetigkeit und irrationale Zahlen,* Vieweg, Braunschweig, Engl. trans, Dover, Ney York, 1963
11. Dedekind, R. (1888), *Was sind und was sollen die Zahlen*? Vieweg, Braunschweig. Engl. Trans, Dover, New York, 1963
12. Fichte, J.G. (1845), *Sämmtliche Werke.* Band 1, Berlin.
13. Frege, G. (1892), *Ueber Sinn und Bedeutung*, Zeitschrift für Philosophie und philosophische Kritik, 100(1892), pp. 25-50.
14. Frege, G. (1976), *Wissenschaftlicher Briefwechsel.* Edited by G. Gabriel, H. Hermes, F. Kambartel, C. Thiel, A. Veraart, Meiner, Hamburg
15. Frege, G. (1884), *Die Grundlagen der Arithmetik. Eine logisch-mathematische Untersuchung über den Begriff der Zahl*, Köbner, Breslau. Engl. trans. Blackwell, Oxford, 1974.
16. Gödel, K. (1930), Die Vollständigkeit der Axiome des logischen Funktionenkalküls, *Monatshefte für Mathematik und Physik* 37, pp. 349–350. Engl. Trans.in K. Gödel, *Collected Works.* 1986–1995, vol. 1 Clarendon Press, Oxford.
17. Gödel, K. (1931), Über formal unentscheidbare Sätze der Principia Mathemtica und verwandter System I, *Monatshefte für Mathematik und Physik* 38, 173–98. Engl. transl. in K. Gödel, *Collected Works.* 1986–1995, vol. 1 Clarendon Press, Oxford, pp. 144–195
18. Henkin, L.(1949), The completeness of the first-order functional calculus, *Journal of Symbolic Logic*, 14, pp.159-166
19. Hilbert, D. (1899), *Die Grundlagen der Geometrie*, Teubner, Leipzig. 10th edition with supplements by P. Bernays, Teubner, Stuttgart, 1968. Engl. transl, by E.T. Townend, Chicago 1902.
20. Hilbert, D. (1918), Axiomatisches Denken, *Mathematische Annalen* 78, pp. 405–415.
21. Hilbert, D. and Bernays, P. (1934–1939), *Grundlagen der Mathematik,* Springer, Berlin, vol. 1 1934, vol. 2 1939.
22. Kronecker, L. (1887), Über den Zahlbegriff, *Journal für die reine und angewandte Mathematik*, 101, 337–355. Repr. in *Werke*, Leipzig, 1899, vol. III/1 249–274.
23. Padoa, A. (1901), Essai d'une théorie algébrique des nombres entiers précédé d'une introduction logique à une théorie déductive quelconque, in *Bibliothèque du Congrès international de philosophie, Paris 1900.* 3. *Logique et histoire des sciences*, Colin, Paris, pp. 309–365.
24. Pasch, M. (1882), *Vorlesungen über neuere Geometrie*, Teubner, Leipzig
25. Peano, G. (1889), *Arithmetices principia nova methodo exposita*, Bocca. Torino.
26. Peano, G. (1889), *I principi di geometria logicamente esposti*, Bocca, Torino
27. Pieri, M. (1899), *Della geometría elementare come sistema ipotetico-deduttivo: monografía del punto e del moto*, in Memorie della Reale Accademia delle Scienze di Torino, 49 seconda serie, pp. 173–222.
28. Pieri, M. (1901), Sur la géométrie envisagée comme un système purement logique, in *Bibliothèque du Congrès International de Philosophie*, Colin, Paris, 1901, pp. 367–404
29. Russell, B. (1901), Recent Work on the Principles of Mathematics, *International Monthly*, 4.
30. Russell, B. (1903), *The Principles of Mathematics*, Cambridge University Press, Cambridge
31. Saccheru, G. (1733), *Euclides ab omni naevo vindicatus…*,Montani, Milano. English trans. by G.B. Halsted, Open Court-Chicago-London, 1920
32. Scholz, H. (1969), *Mathesis Universalis*, Schwabe, Basel-Stuttgart.
33. Sieg, W. (2013), *Hilbert's Program and Beyond,* Oxford University Press, Oxford
34. Whitehead A. N. and Russell, B. (1910–13), *Principia Mathematica*, Cambridge University Press, Cambridge, 3 vols.

Evandro Agazzi is Director of the Interdisciplinary Center for Bioethics of the Panamerican University of Mexico City, and Emeritus Professor of the Universities of Genoa (Italy) and Fribourg (Switzerland). He is Honorary President of the International Academy of Philosophy of Science (Brussel), the International Federation of the Philosophical Societies, the International Institute of Philosophy (Paris). He has published in several languages, as author or editor, more than 90 books and about 1000 papers and book chapters. His main fields of research are logic, philosophy of mathematics, philosophy of physics, general philosophy of science, ethics of science, metaphysics, and bioethics.

Chapter 5
Aristotle's Relations: An Interpretation in Combinatory Logic

Erwin Engeler

Abstract The usual modelling of the syllogisms of the Organon by a calculus of classes does not include relations. Aristotle may however have envisioned them in the first two books as the category of relatives, where he allowed them to compose with themselves. Composition is the main operation in combinatory logic, which therefore offers itself to logicians for a new kind of modelling. The resulting calculus includes also composition of predicates by the logical connectives.

5.1 Introduction

Relations turn up at birthdays,[1] congratulating. Even logicians have them; of the first one, Aristotle, we even know some of their names. But it is a question among historians of axiomatic geometry whether he had the other kind of relations, the ones that modern logicians are concerned with. Indeed, many hold that the Stagirite did not have this concept, and that Greek mathematics, in particular Euclid shows this: the relation of betweenness does not enter his axioms for geometry. The tradition of Aristotelian Logic is often blamed for this serious lacuna. In fact Euclidean axiomatics reached completion only in the 19th century, [1]. Missing the concept of relation is perceived as making it unfit as an adequate logic for developing formal axiomatic mathematics; that had to wait till the Boole-Peano-Russell disruption which eclipsed traditional logic. This, I think, is overstating the case. The Organon itself gives a larger picture of Aristotle's understanding of judgements than those that are formulated by the syllogistics and included relations and functionals. My argument comes in two parts, motivation and formal development. The first part experiments with concepts

[1] The first part of this little essay was written as a gift for my 90th birthday, (see dedication).

E. Engeler (✉)
Eidgenössische Technische Hochschule, Zürich, Switzerland
e-mail: erwin.engeler@math.ethz.ch

F. Ferreira et al. (eds.), *Axiomatic Thinking I*,
https://doi.org/10.1007/978-3-030-77657-2_5

of definitions on an example that Aristotle himself could have handled. These are discussed in terms, motives, that I discern in the Organon, in particular the composition of predicates called "relatives", the use of logical connectives and forms of recursion. Modalities, also an important ingredient of the Organon are not included here, which while feasible are irrelevant to the present theme. In the second part, the compositional aspect of predicates is brought forward and made into the basis for interpreting the *Organon*. This results in the establishment of a logical calculus $\mathcal{E}^{\Lambda}$ of judgements. This provides a model of syllogisms which include relations. The conclusion is, that Aristotle had the means to treat relations but chose not to do so for his syllogistics.

5.2 Aristotle's Relatives

If you'll bear with me, let us see how Aristotle would, and could speak of his relatives, formally, and within the framework of his toolkit, the *Organon*—in my naive and quite ahistoric reading, using an *ad hoc* formalism that we shall later turn into a formal calculus.

5.2.1 Predication

The main grammatical operation is applying a predicate to a subject: $[\mathit{red}] \cdot [\mathit{blood}]$ is a statement which predicates that blood is red. All kinds of thought objects are admitted as predicates; Aristotle divides them into "categories", distinguishing for example between predicates about quantity ("big"), and quality ("red") and relatedness, (the category of "relatives").

Looking at example of relatives, let $[\mathit{mother}]$ predicate of a subject that it is enjoying motherhood. Thus $[\mathit{mother}] \cdot [\mathit{Phastia}]$ states that Phastia is a mother. Nothing prevents us from using this thought object as a predicate:

$$([\mathit{mother}] \cdot [\mathit{Phastia}]) \cdot [\mathit{Aristotle}]$$

tells us that Phastia is the mother of Aristotle. This turn is what Aristotle (Categories, Chap. 7) really meant with the category of relatives; he in fact called this category "things pointing towards something". The above predicate $[\mathit{mother}]$ of motherhood predicates of a subject a that the subject b is her child, "pointing a to b". Thus, a relative predicate in fact introduces a binary relation by composition of predicates. The compositional nature of relatives is an early showing of Currying, a device that became one of the central aspects of combinatory logic.

For the moment, we don't concern ourselves with the question as to what category some predicate or subject might belong, all are treated as relatives; one of the uses that Aristotle gets out of such prescriptions is to avoid predicating nonsense by disqualifying predications between certain categories.

Using the "relative" predications of motherhood and fatherhood, of marriage, and of being male or female applied to family members, we can easily envision the genealogy of Aristotle as a list of such statements. He himself would use many more predicates to talk about his relations: he would use "son", "sister", "grandfather", "sister-in-law", or even "male descendent" etc. as predicative concepts. Let us see how that could fit in.

5.2.2 Explicit Definitions

One way of introducing a new predicative concept is explicitly, as composite predicates. Using variables $a, b, c, \ldots$ for subjects, the definition of b being a child of a is simply

$$([child] \cdot b) \cdot a = ([mother] \cdot a) \cdot b.$$

For some relationships we need logical connectives such as "and" and "or". These are denoted by $\wedge$ and $\vee$, and used on predicates P and Q to obtain $P \wedge Q$ and $P \vee Q$. Thus, the predicate of being a "son" is

$$([son] \cdot b) \cdot a = ([mother] \cdot a) \cdot b \wedge [male]b.$$

Similarly, constituting the predicative concept of a "family": For a, b, c, d to form a core family, predicated by the predicate [*family*] on some individuals a, b, c, d set

$$\begin{aligned}&[family]abcd = \\ &[mother]ac \wedge [mother]ad \wedge [father]bc \wedge [father]bd \wedge [female]a \wedge [male]b.\end{aligned}$$

Notation: we have dropped the center-dot that denotes application and adhere to the convention that sequences of applications are to be understood as parenthesised to the left: uvw is read as $(u \cdot v) \cdot w$, etc.

The idea of "pointing to something" has just been applied again: if P is a relative predication then Pa and Pab may be too. P in the context $Pabc$, for example, would introduce a ternary relation. This leads to a more general notion of explicit definitions:

Formally, an *explicit definition* concerns an expression $\varphi(x_1, x_2, \ldots, x_n)$ built up from variables, predicates (introduced earlier as definienda) and the logical connectives. It defines a new predicate P by the defining equation

$$Px_1x_2 \ldots x_n = \varphi(x_1, x_2, \ldots, x_n).$$

This is the *Principle of Comprehension*. It comprehends the connective structure of φ into a single predicate, an idea that goes back to Schönfinkel and is a basic concept for Curry's combinatory logic.

5.2.3 Implicit Definitions

As ethnologists tell us, all, (even "primitive") cultures allow definitions of familial relatedness, some quite elaborate. The simplest ones are of the explicit kind as above, but this is far from sufficient. Consider the notion of being (maternal) siblings. The desired predicate $[sibling]cd$ hides a mother somewhere in its definition. Aristotle resolves this by introducing the construction "some P".

We denote the construct "some P" by $\varepsilon_x(Px)$. It implies a sort of existential referent. The assertion "some P are Q" would then transcribe to $Q \cdot (\varepsilon_x(Px))$. This device was introduced by Hilbert to be used in his foundational program as a tool for the formal elimination of quantifiers. With it we can define the predicate of being a sibling:

$$[sibling]yz = [mother](\varepsilon_x([mother]xy))z,$$

stating that y's mother x is also mother of z.

5.2.4 Recursion

We now can have uncles, cousins of all kinds, marriages between them, etc., enough to tell the story of Aristotle's relations, and formulate things like "Aristotle is the father-in-law of his niece". However, the concept "male descendent" which was very important at the time is still open. This is a typical case for using a definition by recursion:

$$[mdesc]xy = ([male]y \wedge [father]xy) \vee (\varepsilon_z([father]zy \wedge [mdesc]xz))$$

stating that a male descendent y of x either is a son of x or there is some male descendent who is his father.

Are such definitions admissible? What it amounts to, is that it allows the use of the ε-operator also for predicates as variables: The definiendum $[mdesc]$ is represented by a variable U.

$$[mdesc]xy = \varepsilon_U(([male]y \wedge [father]xy) \vee (\varepsilon_z([father]zy \wedge Uxz))).$$

You may have noticed that the ε operator allows to understand the Aristotelian "some P are Q" as $Q \cdot (\varepsilon_x Px)$ in the present *ad hoc* formalism. We have not used two other basic ingredients of Aristotle's formalism; there was no need of negation and of the companion to "some x are y", namely "all x are y". This will be the task of a later chapter.

5.3 Discussion

5.3.1 *On the* Organon

For all I know, Aristotle would have accepted each definiens in the above definitions as a statement in his sense. But he would have hesitated to call them "categorical" statements. The distinction arises in the course of his development of the *Organon*.[2]

The *Organon*, the way I read it, has the character of a manual, a texbook that instructs the reader in the art of preparing a conclusive argument using well-formed and immediately understandable statements. The *Organon* consists of three parts, "The Categories", "On Interpretation" and "Prior Analytics" (in two books).[3]

The first two introduce the notion of well-formed-ness by a discourse of examples and grammatical distinctions. The question is how to combine predicates while evading non-sensical, ambiguous or misunderstandable compositions as far as possible: How do you compose [*good*], [*man*] and [*shoemaker*] (Int 11)?

You may take conjunction and application to form

$$
\begin{aligned}
&([good] \wedge [shoemaker])[man] \quad \text{or}\\
&([good][shoemaker])[man] \quad \text{or}\\
&[shoemaker]([good][man]) \quad \text{or}\\
&([good][man])[shoemaker],
\end{aligned}
$$

each with a valid and different meaning.

Disjunction can enter with predicates of the category of qualities such as colour;

$$
\begin{aligned}
&([black] \vee [white])[man], \quad \text{(Cat 8), or}\\
&([green] \vee [blue])[turquoise], \quad \text{or}\\
&[green]([blue][turquoise]) \quad \text{or}\\
&([blue][green])[turquoise],
\end{aligned}
$$

as you may judge the quality of the stone.

Negation enters in two pairs of *opposites* (Cat 10, Int 6–10):

---

[2] Consulted translations: Owen [2] and Smith [3].

[3] These are cited as (Cat), (Int), (PA A), (PA B) with only the chapters indicated, e.g. (Cat 8) is Chap. 8 in Categories.

"some P are Q" as against "no P is Q", and
"all P are Q" as against "some P are not Q".

These are the four kinds of statements that Aristotle calls *categorical statements*.

In Prior Analytics, Aristotle gets down to the business of constructing conclusive sequences of arguments. Already required for categorical statements, each argument should be terse and immediately understandable. He chose to restrict the form of statements to the four categorical ones above, which for convenience we denote in the form $\varphi(x, y)$, $\psi(x, y)$, etc.

And then he shows *unguem leonis*[4]: he develops a formal system of logic based on logical arguments, called *syllogism*, of the form

$$\varphi(P, Q), \psi(Q, R) \vdash \chi(P, R),$$

expressing that from the categorical statements $\varphi(P, Q)$ and $\psi(Q, R)$ you may infer $\chi(P, R)$, (PA A 1–14). Since there are four of these, there is a plethora of such deductive patterns. Aristotle proceeds to eliminate all but fourteen of them by showing, using counterexamples, which of them preserve the truth of the statements.

The high point of Prior Analytics is the proof of a metatheorem: all fourteen can be deduced from just two of them, (PA A 23).

5.3.2 *Critiques of the* Organon

By this metatheorem, Aristotle establishes a sort of completeness for his syllogistic proof system. This is different from the present notion of completeness in mathematical logic which involves models.

The story of *models* for Aristotle's logic can be traced from Boole's Laws of Thoughts, to Łukasiewicz [4], Shepherdson [5] and Corcoran [6] in the 20th century. The view developed that this logic dealt with unary truth-functions which could be understood as classes. To accomodate the interpretation of categorical sentences, the classes had to be non-empty. Although he would not put it this way, Corcoran, for example, interpreted these as:

"all A are B" is: $A \subseteq B$,
"no A is B" is: $A \cap B = \emptyset$,
"some A are B" is: $A \cap B \neq \emptyset$,
"some A are not B" is: $A \nsubseteq B$.

With such interpretations, syllogistic may be treated as a calculus of equations [5], something that Boole seems to have had in mind. It reflects a rather impoverished *Organon*. But if people considered this sort of model as the true interpretation of Aris-

[4] "Writing with the the claws of a lion.".

totle, there is no place for relations, and the opinion that he therefore is responsible for the lacunae of Euclidean axiomatics gets some support.

Scholars of Euclidean axiomatics such as de Risi [7], do not share this opinion: Aristotle, an alumnus of Plato's academy, where famously nobody entered without it, did know geometry. Prior Analytics contains a geometric proof of the equality of base angles in an equilateral triangle, (PA A 26), and "the principle of Aristotle" of Euclidean tradition is related to parallelity.

Posterior Analytics (Chap. 19), reflects Aristotle's understanding of *recursion* as a mentally completed inductive definition of a concept. "Mental completion" is hard to understand without a set-theoretic mindset, and it was a controversial issue for many commentators of Aristotle.[5]

Logic definitely turned away from the Aristotelian tradition only at the turn to the 20th century. Bertrand Russell was an important mover in this. He had learned classical logic as a student but also had read up on Leibniz's attempts at reforming logic, critically. He asked himself several questions inspired by this reading, in particular one that is relevant here:

> "(1) Are all propositions reducible to the subject-predicate form?" [8, p. 13][6]

On the following pages of this book he proceeds to demonstrate by examples, that a logic adequate for mathematics cannot dispense with relations. Indeed later, in *Principia Mathematica* they are a central ingredient.

5.4 The *Organon* in Combinatory Logic

An adequate mathematical model for more of Aristotle's logic seems to be missing. This section describes my attempt to construct one.

Logicians who have followed me to this place have long noticed that the *ad hoc* formalism that I introduced in the discussion of Aristotle's family is in fact an extension of combinatory logic. Predicates, logical concepts and operations were added to the language by defining equations in these terms, combined with the combinatory application operation. This will now be turned into a formal calculus based on a language $\mathcal{E}$ which extends the language of combinatory logic.

Before introducing this language, we choose a mathematical structure into which the language will be interpreted. Because it includes combinatory logic, we need a model for that.

---

[5] It was therefore only settled after the creation of set theory.

[6] I'm grateful to Prof. V. de Risi for pointing me to this book in recent correspondence.

5.4.1 The Modelling Structure

The Language $\mathcal{E}$ of the modelling extends that of combinatory logic.

We first consider a fragment $\mathcal{E}_0$ of $\mathcal{E}$. It consists of expressions, built from variables and constant predicates by the operations of application, the epsilon operator ε_x and the alpha operator α_x for variables x. These operations bind the variable x. We shall distinguish between objects that are called *predicates* and predications. A *predication* may be obtained by applying a predicate P to a variable x, "predicating something about x", written $P \cdot x$.

The basis for our interpretation of the language $\mathcal{E}_0$ is the graph-model of combinatory logic, Scott (1969), Plotkin (1972), Engeler (1981).

Let A be a non-empty set and define recursively

$$G_0(A) = A,$$
$$G_{n+1}(A) = A \cup \{\alpha \to a : \alpha \text{ finite or empty}, \alpha \subseteq G_n(A), a \in G(A)\},$$

where $\alpha \to a$ is a notation for the pair $\langle \alpha, a \rangle$.

The union of these $G_n(A)$ is denoted by $G(A)$.

The combinatory application operation is defined on subsets M and N of $G(A)$ as

$$M \cdot N = \{x : \alpha \to x \in M, \alpha \subseteq N\}.$$

With this interpretation of the application operation, the set $G(A)$ can be shown to be a model of combinatory logic. The elements of the model are the subsets of $G(A)$. We shall show below that the model satisfies the *Comprehension Axiom of Combinatory Logic*:

For every expression $\varphi(x_1, \ldots, x_n)$ built up by the application operation from the constants and variables (interpreted as subsets of $G(A)$), there exists an element M of the model such that

$$M \cdot x_1 \cdot x_2 \ldots \cdot x_n = \varphi(x_1, \ldots, x_n).$$

The proof of this theorem actually produces an algorithm of comprehension to obtain M. Observe that all elements of M have the form

$$\alpha_1 \to (\alpha_2 \to \ldots (\alpha_n \to a)) \text{ with } \alpha_i \subseteq G(A) \text{ finite or empty, and } a \in G(A).$$

For the next steps we shall rely on *Schönfinkel's comprehension theorem*. He shows that if we have the "combinators" $\boldsymbol{S}$ and $\boldsymbol{K}$ for which the equations $\boldsymbol{S}PQR = PQ(PR)$ and $\boldsymbol{K}PQ = Q$ hold for all predicates P, Q, R, then there is the following conversion:

COMPREHENSION THEOREM OF COMBINATORY LOGIC. For every combinatory expression $\varphi(x_1, \ldots x_n)$ built up from constants and the variables x_i there is a purely applicative expression $\psi(\boldsymbol{S}, \boldsymbol{K})$ such that

$$((((\psi(\boldsymbol{S}, \boldsymbol{K}) \cdot x_1) \cdot x_2) \cdots) x_n = \varphi(x_1, \ldots, x_n).$$

For completing the proof we need only produce interpretations of the two constants and show, by inspection, that these conform to the equations, and thereby verify that by our interpretation we have in fact a model of combinatory logic. This is done in this author's 1981 paper on graph models, [9]. Here are the interpretations of the two combinators:

$$[\boldsymbol{K}] = \{\{a\} \to (\emptyset \to a) : a \in G(A)\},$$
$$[\boldsymbol{S}] = \{(\{\tau \to (\{r_1, \ldots, r_n\} \to s)\} \to (\{\sigma_1 \to r_1, \ldots \sigma_n \to r_n\} \to (\sigma \to s))): \\ n \geq 0, r_1, \ldots, r_n \in G(A), \tau \cup \sigma_1 \cup \cdots \cup \sigma_n = \sigma \subseteq G(A), \sigma \text{ finite}\}.$$

5.4.2 The Combinatory Interpretation of Categorical Predication

Combinatory Predicates are composed by the operation of application from predicate constants C_j and variables x_i to form expressions $\varphi(C_1, \ldots, C_m, x_1, \ldots x_n)$. The constants C_j are interpreted as subsets $[C_j]$ of $G(A)$, each variable x_i ranges over a specific subset of $G(A)$. Their mention in φ is usually suppressed.

Some predicates can be used for *predications*: If the predicate P is interpreted as $[P]$, a subset of $G(A)$, and $[P]$ is a set of elements of $G(A)$ of the form

$$(\alpha_1 \to (\alpha_2 \to \ldots (\alpha_n \to a)) \\ \text{with } a \in G(A), \alpha_i \subseteq G(A) \text{ finite or empty}, i = 1, \ldots, n,$$

then $[P]$ can act as a predication $[P] \cdot [x_1] \cdots [x_n]$ on these variables, interpreted as subsets $[x_i]$ of $G(A)$.

Notation: Where no ambiguity results we may omit the brackets on interpreted variables in the future.

The *intuition behind this interpretation* of predication is that $[P]$ as a predication expresses some *facts* about each subject-variable x_i. These facts are the extent to which x_i conforms to the predicate P, the conformity being expressed by the corresponding sets α_i. We call these facts "*attributes*".

An interpretation of a predication is perhaps best illustrated by an example which we take from the family context of Sect. 5.2. The interpretation of the parent predicate $[parent(x, y)]$ is a set of expressions $(\alpha_1 \to (\alpha_2 \to a))$ with $\alpha_1, \alpha_2 \subseteq G(A)$, $a \in A$, and where A is a set of people, each is present with the individual attributes.—In distinction to Sect. 5.2 we added the variables inside the the brackets for clarity, they relate the variables x, y to the sets α_1, α_2 in that order. Each set α_i is understood as a set of attributes: α_1 of being a parent, α_2 of being a child.

The meaning of the predication $[parent(x, y)] \cdot [x] \cdot [y]$ therefore is: "$[y]$ is the set of people in A for whom $[x]$ is a set of parents". Specifically: α_1 is the set of expressions $\{x\} \to (\{z\} \to z)$ for x male, $\{y\} \to (\{z\} \to z)$ for female, and α_2 consists of all $\{z\} \to z$ for the children z, with $x, y, z \in A$. The predication produces the children of x and y if x is male and y female.

Categorical Predicates, the analog to the categorical statements in the *Organon*, arise from combinatory predicates by using "*for some*" and "*for all*", referring to the variables of a predicate. They constitute our language $\mathcal{E}_0$. We use $\varepsilon_x\varphi(x)$ to denote "some x has $\varphi(x)$" and $\alpha_x\varphi(x)$ to denote "all x have $\varphi(x)$".

Extending the modelling to the ε-operator is a bit subtle. The term $\varepsilon_x\varphi(x)$ is to be interpreted as the result of a recursion in the sense of "completed induction". Recall that the modelling of the language $\mathcal{E}_0$ is a process of finding denotations for elements of the language in a combinatory model. The modelling of $\varepsilon_x\varphi(x)$ involves the determination of an object F, a subset of $G(A)$, which has the property $\varphi(F)$. "Recursion" means that such an object is already determined by an object F_0, the basis of recursion. This implies that the process of interpretation calls here for the choice of a particular object F_0, which, as the case may be, is a challenge for the ingenuity of the modeller.

Given a unary predicate P, the object $[\varepsilon_x(Px)]$ is therefore determined by the interpretation, which proposes an initial set $F_0 \subseteq G(A)$ with the property $F_0 \subseteq [P] \cdot F_0$, and yields

$$[\varepsilon_x(Px)] = \bigcup_n [P]^n F_0 = F, \quad [P]^n \text{ denoting the } n\text{-th iteration.}$$

Then F is a fixpoint of $[P]$, noting $[P] \cdot F = F$.

The finding of an appropriate F_0 is the cardinal point on which it turns whether or not the interpretation of the predication becomes vacuous, (see e.g. Sect. 5.5.1 below on the existence of a model for projective geometry). F_0 always exists, determined by the interpretation, in the worst case it is $F_0 = F = \emptyset$.

The α-operator is interpreted as

$$[\alpha_x(Px)] = [P] \cdot ext_{[P]}([x]),$$

where $ext_{[P]}([x]) = \{a : \exists\alpha \to a \in [P]\}$ is the set of possible values for $[P]x$.

The Aristotelian "some P are Q" thus translates into $[Q] \cdot [\varepsilon_x(Px)]$ and "all P are Q" into $[Q] \cdot [\alpha_x(Px)]$.

To extend these operations to n-ary predications we make another use of comprehension to separate out a specific variable in an expression $\varphi(x_1, \ldots, x_n)$:

$$(\varphi_j(x_1, \ldots, x_n) \cdot x_1 \cdots x_{j-1} \cdot x_{j+1} \cdots x_n) \cdot x_j = \varphi(x_1, \ldots, x_n) \cdot x_1 \cdots x_n.$$

The ε-operator and α-operators for n-ary predicates $[\varphi]$ are defined accordingly:

$$[\varepsilon_{x_j}(\varphi_j(x_1, \ldots, x_n))] \cdot x_1 \cdots x_{j-1} \cdot x_{j+1} \cdots x_n = F,$$

where, for F_0 given by the interpretation,

$$F = \bigcup_m ([\varphi_j(x_1, \ldots, x_n)] \cdot x_1 \cdots x_{j-1} \cdot x_{j+1} \cdots x_n)^m \cdot F_0.$$

F is a set function with $n-1$ variables.

$$[\alpha_{x_j}([\varphi_j(x_1, \ldots, x_n))]] \cdot x_1 \cdots x_{j-1} \cdot x_{j+1} \cdots x_n = \\ [\varphi_j(x_1, \ldots, x_n)] \cdot x_1 \cdots x_{j-1} \cdot x_{j+1} \cdots x_n \cdot ext_{[\varphi_j]}([x_j]),$$

where

$$ext_{[\varphi_j]}([x_j]) = \{a : \exists(\alpha_1 \to \cdots \to (\alpha_n \to a)) \in [\varphi_j(x_1, \ldots, x_n))]\}.$$

REMARK. Two categorical statements "no P is Q" and "some P are not Q" are missing in $\mathcal{E}_0$. They are added in the next section in the context of negation. This is an expository choice. In fact they could have been added here separately, which would make $\mathcal{E}_0$ the full categorical language.

5.4.3 The Interpretation of Logical Connectives and Truth

Predications as defined above are "factual" interpretations, they produce a set of facts $[P] \cdot x_1 \cdots x_n$. Our interpretation of the language $\mathcal{E}_0$ resulted in a calculus of facts and as such cannot really be called a logical calculus. It lacks the logical connectives and judgements about the truth of a predication.

Predications lend themselves to logical composition by the connectives $\wedge$, $\vee$ and $\neg$. These constitute a language extension $\mathcal{E}$ of $\mathcal{E}_0$. The interpretation is extended to $\mathcal{E}$ recursively on the structure of the logical composition: the evaluation of

$$[\varphi(x_1, \ldots, x_n)]x_1 \cdots x_n \wedge [\psi(x_1, \ldots, x_n)]x_1 \cdots x_n$$

is

$$[\varphi(x_1, \ldots, x_n)]x_1 \cdots x_n \cap [\psi(x_1, \ldots, x_n)]x_1 \cdots x_n,$$

correspondingly with $\vee$ and $\cup$, where we conformed the two predications to combined variables $x_1, \ldots, x_n$ by comprehension.

For negation we set

$$[\neg\varphi(x_1,\ldots,x_n)]x_1\cdots x_n = ext_{[\varphi_n]}([x_n]) - [\varphi(x_1,\ldots,x_n)]x_1\cdots x_n.$$

This concludes the definition of the logical predications. In particular, we can now express all the syllogistic statements "some P are Q", ..., "some P are not Q", and our *ad hoc* formalism in Sect. 5.2 is thereby legitimised.

OBSERVATION. The operation of negation could have been added separately in the definition of the language $\mathcal{E}_0$ which would make it possible to add to it the two missing categorical statements "no P is Q" and "some P are not Q". The extended language $\mathcal{E}$ thus includes the full Aristotelian language of categorical statements.

The intuition behind our *Truth Definition* for a unary predicate P, modelled by a set of expressions $\alpha_i \to a$ is that $[P]x$ is true if x has all the attributes that are required by P, that is $[P]x = \{a : \alpha \to a \in [P]\}$. Correspondingly the truth definition for arbitrary predications in $\mathcal{E}$ is

$[\varphi(x_1,\ldots,x_n)]x_1\cdots x_n$ is $true$, denoted by $\top$,
 if $[\varphi_j(x_1,\ldots,x_n)]x_1\cdots x_{j-1}\cdot x_{j+1}\cdots x_n\cdot x_j = ext_{[\varphi_j]}([x_j])$ for each j.
$[\varphi(x_1,\ldots,x_n)x_1\cdots x_n$ is $false$, denoted by $\bot$,
 if it is a proper non-empty subset of $ext_{[\varphi_j]}([x_j])$ for some j.

In all other cases it is *indeterminate*, denoted by $\triangle$.

5.4.4 Relational Predications: From Factual to Logical Interpretation

The "facts" produced by predications $[P]\cdot x_i\cdots x_n$ in $\mathcal{E}$ do not reflect the actual relations between the arguments x_i. This can be accomplished by introducing *relational predications* $[P]_R$: If $[P]$ consists of elements$(\alpha_1 \to \cdots \to (\alpha_n \to a))$ with $a \in G(A)$, $\alpha_i \subseteq G(A)$ finite and nonempty, then

$$[P]_R\cdot x_1\cdots x_n \text{ is the set of all } \langle a_1,\ldots,a_n\rangle : \exists(\alpha_1 \to \cdots \to (\alpha_n \to a)) \in [P]$$

such that $a_i \subseteq x_i \subseteq \alpha_i$, $i = 1,\ldots,n$. The subscript R is tacitly understood in the following.

Let $\mathcal{E}$ be extended to $\mathcal{E}^\Lambda(C_1,\ldots,C_m)$ by introducing a valuation of the relational predicate constants $C_1,\ldots,C_m$ of $\mathcal{E}$.

The *factual interpretation* of relational predications distinguishes items that verify or falsify it. This distinction is based on a valuation Λ_i for each one of the constant predicates $[C_i]$. We denote the valuation of a tuple $\langle a_1,\ldots,a_n\rangle$ by $\langle a_1,\ldots,a_n\rangle^{\Lambda_i} = \langle a_1,\ldots,a_n\rangle^\top$ if it is a *verifying* fact, by $\langle a_1,\ldots,a_n\rangle^\bot$ if it is *falsifying*.

For n-ary predicate constants $[C_i]$ the facts that are valued by the valuation Λ_i are the tuples $\langle a_1, \ldots, a_n \rangle \in D_i$, where

$$D_i = \{\langle a_1, \ldots, a_n \rangle : (\alpha_1 \to \ldots (\alpha_{n-1} \to a_j)) \in [(C_i)_j], j = 1, \ldots, n\}.$$

We define the corresponding predication $[C_i]^{\Lambda_i}$ by the set of objects

$$(\alpha_1 \to \ldots (\alpha_n \to \langle a_1, \ldots, a_n \rangle))^{\Lambda_i}$$

where $(\alpha_1 \to (\alpha_2 \to \cdots \to (\alpha_n \to a_n))) \in [C_i]$, $\langle a_1, \ldots, a_n \rangle \in D_i$.

As a result, the factual interpretation of the relational predication C_i is the set function $[C_i]^{\Lambda_i} \cdot x_1 \cdots x_n$ which produces a set containing elements $\langle a_1, \ldots, a_n \rangle^\top$, $\langle a_1, \ldots, a_n \rangle^\perp$ and $\triangle$, (for the cases where the predication returns the empty set on the given inputs $x_1, \ldots, x_n$).

The factual interpretation of the language $\mathcal{E}^\Lambda(C_1, \ldots, C_m)$ is based on the valuation Λ_i for each C_i, and then extended over logical composition, the ε- and α-operations and the quantifiers $\exists$ and $\forall$ as follows: The valuation Λ maps each predication into a set of the above objects. We represent each such set as a propositional formula consisting of the valued tuples $\langle a_1, \ldots, a_n \rangle$. The valuation Λ assigns to each of them the truth-value *true* or *false* as indicated by the superscripts. To obtain the truth-value of a constant predication, the set produced by it is interpreted as the conjunction of these elements as a formula in a propositional logic. The presence of $\triangle$ in a propositional formula assigns to it the value *indeterminate*.

This interpretation is then extended as follows to predications obtained by the logical operations:

The conjunction and disjunction of predications $[\varphi(x_1, \ldots x_n)]^\Lambda \cdot x_1 \cdots x_n$ and $[\psi(y_1, \ldots y_m)]^\Lambda \cdot y_1 \cdots y_m$ are conjunctions respectively disjunctions of the corresponding propositional expressions. The interpretation of negation is obtained by inverting all $\top$ to $\perp$ and all $\perp$ to $\top$ in the valuations of the tuples $\langle a_1, \ldots, a_n \rangle^\top$, $\langle a_1, \ldots, a_n \rangle^\perp$.

The factual interpretation of the ε- and α-operators on a predicate $\varphi(x_1, \ldots, x_n)$ is obtained as

$$\begin{aligned}
&[\varepsilon_{x_j}(\varphi(x_1, \ldots, x_n))]^\Lambda \cdot x_1 \cdots x_{j-1} \cdot x_{j+1} \cdots x_n = \\
&\quad \{\langle a_1, \ldots, a_{j-1}, a_{j+1}, \ldots, a_n, a_j \rangle^\top : \\
&\qquad a_j \in [\varepsilon_{x_j}(\varphi_j(x_1, \ldots, x_n))]^\Lambda \cdot x_1 \cdots x_{j-1} \cdot x_{j+1} \cdots x_n \cdot x_j\}, \\
&[\alpha_{x_j}(\varphi(x_1, \ldots, x_n))]^\Lambda \cdot x_1 \cdots x_{j-1} \cdot x_{j+1} \cdots x_n = \\
&\quad \{\langle a_1, \ldots, a_{j-1}, a_{j+1}, \ldots, a_n, a_j \rangle^\top : \\
&\qquad a_j \in [\varphi_j(x_1, \ldots, x_n)]^\Lambda \cdot x_1 \cdots x_{j-1} \cdot x_{j+1} \cdots x_n \cdot ext_{[\varphi_j]}([x_j])\} \\
&\quad \cup \{\langle a_1, \ldots, a_{j-1}, a_{j+1}, \ldots, a_n, a_j \rangle^\perp : \\
&\qquad\quad a_j \in [\neg\varphi_j(x_1, \ldots, x_n)]^\Lambda \cdot x_1 \cdots x_{j-1} \cdot x_{j+1} \cdots x_n \cdot ext_{[\varphi_j]}([x_j])\}.
\end{aligned}$$

Existential and universal quantifiers can now be introduced as follows:

$$[\exists x_j \varphi(x_1, \ldots, x_n)]^\Lambda \cdot x_1 \cdots x_{j-1} \cdot x_{j+1} \cdots x_n =$$
$$[\varepsilon_{x_j}(\varphi(x_1, \ldots, x_n))]^\Lambda \cdot x_1 \cdots x_{j-1} \cdot x_{j+1} \cdots x_n,$$
$$[\forall x_j \varphi_j(x_1, \ldots, x_n)]^\Lambda \cdot x_1 \cdots x_{j-1} \cdot x_{j+1} \cdots x_n =$$
$$[\alpha_{x_j}(\varphi(x_1, \ldots, x_n))]^\Lambda \cdot x_1 \cdots x_{j-1} \cdot x_{j+1} \cdots x_n.$$

Altogether, we have now extended the factual interpretation of predications to all of the language $\mathcal{E}^\Lambda(C_1, \ldots, C_m)$. Each predication produces a propositional formula containing elements of the form $\langle a_1, \ldots, a_n\rangle^\perp$, $\langle a_1, \ldots, a_n\rangle^\top$ and Δ. This completes the factual interpretation of $\mathcal{E}^\Lambda(C_1, \ldots, C_m)$; it assigns a propositional formula of verifying, falsifying tuples and Δ-s.

The *truth-value* of the predication is obtained by evaluating the tuples of the propositional components as true, false and indeterminate as above.

5.5 Discussion

5.5.1 Projective Geometry

Looking for an example to discuss our interpretation $\mathcal{E}$ and $\mathcal{E}^\Lambda$ of the *Organon*, recall Plato's advice in the *Republic*, (Chap. 7), that with geometry you can educate the mind.

We take "geometry" here as a first-order theory in mathematical logic, and for simplicity restrict to projective geometry.

The first-order models in mathematical logic are relational structures $\langle A, R_1, \ldots, R_n, c_1, \ldots, c_n\rangle$ with relations $R_i \subseteq A^{k_i}$ and individual constants $c_j \in A$. In our modelling the relations R_i correspond to combinatory constants that denote k_i-ary predicates; the constants c_j also correspond to combinatory constants. These objects are then interpreted as subsets of $G(A)$ according to the above definitions.

As an example, consider a mathematical structure such as a projective plane, understood as a set P of "points", L as a set of "lines" with the binary relation of "incidence" $Inc \subseteq P \times L$, where $P \cap L = \emptyset$. These are subject to some axioms such as: "For any two points there is a unique line on which they are incident."

The combinatory model is based on this A for the construction of $G(A)$. In $\mathcal{E}$ the incidence relation is interpreted as

$$[Inc]^\Lambda = \{(\{p\} \to (\{p\} \to l)) \to (\{p\} \to l) : p \in P, l \in L, \langle p, l\rangle \in Inc\}.$$

In $\mathcal{E}^\Lambda(Inc)$ it would be

$$[Inc] = \{(\{p\} \to (\{p\} \to l)) \to (\{p\} \to \langle p, l\rangle) : p \in P, l \in L, \langle p, l\rangle \in Inc\}$$

An equality predicate is needed here only for points and lines and can therefore be viewed as a binary predicate constant with the interpretation

$$[eq] = \{(\{x\} \to (\{y\} \to y)) \to (\{y\} \to y) : x = y, x, y \in P \cup L\} \text{ in } \mathcal{E},$$

with the corresponding relational predication

$$[eq]^{\wedge} = \{(\{x\} \to (\{y\} \to y)) \to (\{x\} \to \langle x, y\rangle) : x = y, x, y \in P \cup L\} \text{ in } \mathcal{E}^{\wedge}.$$

Using the quantifiers introduced earlier, the above axiom is interpreted as

$$\forall x_1 \forall x_2([eq]^{\wedge} x_1 x_2 \vee \\ \exists y([Inc]^{\wedge} x_1 y \wedge [Inc]^{\wedge} x_2 y \wedge \forall z([\neg Inc]^{\wedge} x_1 z \vee [\neg Inc]^{\wedge} x_2 z \vee [eq]^{\wedge} yz))).$$

with the parameter [*Inc*].—The other axioms would be represented in the same fashion and combined into a logical predicate denoted by $\pi(Inc)$. It then turns into an exercise of inventiveness in finding the required fixpoints, and of formal persistence to verify that it returns the value *true* on some given model of projective plane geometry, that is for a given binary relation of incidence, e.g. the *Fano plane* of seven points and seven lines.

Of course, projective geometry itself may be considered as a defined predicate, obtained by using a suitably defined recursion on a predication: Let the variable X be substituted for *Inc* in the axiom-expression $\pi(Inc)$. The recursion equation $X = \varepsilon_X \pi(X))$ defines the geometric concept of a projective plane. The modelling of this formula starts with fixing on a given set of points and lines, these may come from definitions in terms of finite fields or $\mathbb{Q}$, $\mathbb{R}$ or $\mathbb{C}$. Here again the modelling consists in the finding of a fixpoint, which is the crucial matter in obtaining an actual model. If the language has more than one predicate constant, e.g. one for betweenness, one would of course use joint recursion for incidence and betweenness.

REMARK. Perhaps it is worth mentioning that this use of the predicate $\pi(X)$ is an example of introducing additional predicate constants with which one may extend the language to encompass additional concepts. This corresponds to the familiar way to define new mathematical concepts and structures.

5.5.2 *Facts and Thoughts*

The restriction of "facts" to the basic set A as the extension of the variables is natural in simple contexts like the Aristotle family. We also used it in the context of projective geometry treating of "facts" about points and lines. But this is too restrictive even in this case: Consider the notion of point-functions, for example projectivities. The object f is a point-function if $fpq_1 = fpq_2$ for all $q_1, q_2 \in P$. This can be expressed by a predication [*fun*] defined by

$$[fun]f = [eq] \cdot \alpha_{q_1}(fpq_1) \cdot \alpha_{q_2}(fpq_2),$$

using the predication $[eq]$ from above. Observe that $[fun]$ is an element of $G_2(A)$.

The ε-operator also creates "facts": Consider the line connecting two points p_1, p_2, expressed by $\varepsilon_l([eq] \cdot [Inc](p_1 l) \cdot (p_2 l))$, which is a function on p_1, p_2, called a *Skolem-function* in logic. It is an element of $G_2(A)$.

A "Skolem-function" f is thus the result of a recursion, a concept that we traced back to Aristotle's notion of a mentally completed definition of induction. It is therefore legitimate to call this object a *thought*.—Anyway, I would have preferred "*thoughts*" over "*predicates*" and "*predications*". The latter are naturals in the Aristotelian context. But I see predicates as thoughts, as sets or patterns of small and big notions: The predication $[P] \cdot x$ is perceived as applying a thought P to a thought x, checking to what extent the thought $[P]$ applies to x. This perception is the background of my modelling and is connected to my work on neural algebra which treats of thoughts as patterns of firing neurons.

5.5.3 Algorithmic Logic

Let me not forget my own brain-child, algorithmic logic, [1], now approaching retirement age after a long career. It treats of algorithmic properties of structures. Its predicates are of the form $\pi(x_1, \ldots, x_n)$ which denotes a program π with the input variables $x_1, \ldots x_n$ of elements of the structure. Since combinatory logic can deal with the notions of computation and termination, there is an important point of contact here which merits elaboration.

Programs $\pi(x_1, \ldots, x_n)$ are composed of individual instructions, namely assignments of the form $z := f(x, y)$, decisions such as $x < y$. These correspond, loosely speaking, to our predicate constants. Program statements are composed by successive execution $(\pi_1(x_1, \ldots, x_n)); (\pi_2(x_1, \ldots, x_n))$ and recursion. These essentially correspond to composition and the ε-operation. Finally, the factual interpretation $[\pi(x_1, \ldots, x_n)]^\Lambda \cdot x_1 \cdots x_n$ of a program is the so-called denotational semantics of the program. The valuation Λ is understood as the valuation of tuples of elements of the relations and functions in some relational structure. The program statement $[\pi(x_1, \ldots, x_n)]^\Lambda x_1 \cdots x_n$ evaluates to the result of executing the program on the input assignment.

For the logic of programs, the "*algorithmic logic*", we chose to evaluate a program as "true" in a relational structure if it halts on all inputs.

5.6 Conclusions and Scholia

5.6.1 On the Discussion of Aristotle's Relations

This discussion may be conducted in the interpretation of the language $\mathcal{E}^{\Lambda}$ in Sect. 5.4.4 above which captures, extends and completes the *ad hoc* formalism used in our motivational Sect. 5.2. The example fits nicely into this framework: The set A lists the names of the members of the family. The binary relations for motherhood, fatherhood and marriage are lists of pairs of names, and singletons for being male or female. Thus, the interpretation $[father]^{\Lambda}$ for fatherhood is a case of binary relations which are represented in the form

$$\{\{x\} \to (\{y\} \to y)) \to (\{x\} \to \langle x, y \rangle).$$

Therefore fatherhood in the Aristotle family would be represented by a set containing the substitution instances

$$\begin{aligned} x &:= Niarchus1,\ y := Aristotle \quad \text{and} \\ x &:= Niarchus1,\ y := Arimnestus \quad \text{as well as} \\ x &:= Aristotle,\ y := Niarchus2. \end{aligned}$$

A more adequate modelling of fatherhood would need more attributes from the vital statistics of the family members, e.g. profession, place and date of birth, etc. to exclude false claims of fatherhood. Above, we have artificially distinguished the two people called "Niarchus" by adding the distinction to the names.

In this setting it becomes clear that Aristotle had the conceptual means to conceive of, and formally treat, relations. But the primary goal of the *Organon* did not require this.

This concludes our search for Aristotle's relations. The missing relations of Euclid, a blemish on his axioms, were not noticed at the time because geometry was understood by Aristotle as describing geometric properties of lines and other objects unquestioned as continuous. As pointed out by de Risi in [10] this was only put into question in the 16th century and not fully received into the understanding of space till the 19th century.[7]

[7] The de Risi references were pointed out to me by Prof. M. Beeson.

5.6.2 Scholia

What have we learned from our combinatory experiment on the *Organon*?

I: The objects of syllogisms, the categorical statements, were interpreted as elements of the language $\mathcal{E}$ which expands the set of terms of combinatory logic. This combinatory interpretation thus becomes a model of syllogistics comparable to the models of Łucasiewicz and others.

II: The categorical statements themselves are statements about properties of facts, traditionally about individual facts. This aspect is captured by our language $\mathcal{E}$.

III: The logical interpretation of the language $\mathcal{E}^{\Lambda}(C_1, \ldots, C_m)$ is based on facts about relations as n-ary predications. This is probably the most adequate rendering of my understanding of an extension of the *Organon* if it were to include relations.

IV: Logical interpretations of predications turn them into *Judgements*. Our distinction between factual and logical interpretation recalls the famous distinction between judgements *de re* and judgements *de dicto* going back to 12th century Scholastics, when Peter Abelard derived *de dicto* from *de re*, as we do.[8]

V: Our modelling addressed the *semantics* of $\mathcal{E}$ and $\mathcal{E}^{\Lambda}$ and not its *deductions*. But this is another chapter.

5.6.3 Apologia and Dedication

Who knows how Aristotle would react to my experiment. I picture him and Euclid as little Raffael-angels looking down on our travails with mischievous interest.

My interpretation is based on my individual reading of his *Organon*. Individual does not mean indivisible; I may have had two minds about a number of things. Also, while I avoided the pitfalls of turning the logical connectives into predicates and moreover turn syllogisms themselves into statements, I do not propose to justify this here.

Finally, I claim the privilege of a very old man and refrain from the labours of performing all the verifications and of following up my own suggestions. Let the friendly reader smile and forgive. Here, thanks are due to Michael Beeson whose detailed comments on the manuscript were very helpful.

I dedicate this little essay to all my friends and students who troubled themselves to remember me and my birthday at the symposium in Zurich, early 2020. And to all those that could not be present because of the distances that destiny put between us, in particular to my late friend and colleague Ernst Specker whom this symposium was meant to honour too. Special mention goes to the three organisers Gerhard Jäger,

[8] This was strongly contested at the time by Bernard de Clairvaux, who maintained that the dogmata of the Church (to which Abelard addressed himself) were *de dicto* statements and not to be made dependent on judgements *de re*. Bernard was sainted, Abelard not.

Reinhard Kahle and Giovanni Sommaruga, the sponsors that they found, and to my *alma mater*, the ETH, for its hospitality.

References

1. Erwin Engeler. Algorithmic properties of structures. *Mathematical Systems Theory*, 1:183–193, 1967.
2. Octavius Freire Owen. *The Organon, or Logical Treatisies of Aristotle with an Introduction to Porphyry*. Henry G. Bohn, London, 1853. (Two volumes).
3. Robin Smith. *Prior Analytics*. Hackett Publ.Co., Indianapolis Cambridge, 1989.
4. Jan Łukasiewicz. *Aristotle's Syllogistics from the Standpoint of Modern Logic*. Clarendon Press, Oxford, second edition, 1957.
5. John C. Shepherdson. On the interpretation of aristotelean syllogistics. *Journal of Symbolic Logic*, 21:137–147, 1956.
6. John Corcoran. Aristotle's prior analytics and Boole's laws of thoughts. *History and Philosophy of Logic*, 24:261–288, 2003.
7. Vincenzo de Risi. The development of euclidean axiomatics. *Archiv for the History of the Exact Sciences*, 70:561–676, 2016.
8. Bertrand Russell. *A Critical Exposition of the Philosophy of Leibniz*. Cambridge Univ. Press, Cambridge, 1900.
9. Erwin Engeler. Algebras and combinators. *Algebra Universalis*, 13:389–392, 1981.
10. Vincenzo de Risi. Francesco Patrizi and the new geometry of space. In Koen Vermeir and Jonathan Regier, editors, *Boundaries, Extents and Circulations: Space and Spaciality in Early Modern Natural Philosophy*, volume 41 of *Studies in History and Philosophy of Science*, chapter 3, pages 55–100. Springer International Publishing, Switzerland, 2016a.

Erwin Engeler born 1930, obtained his Dr. sc. math. at the ETH in 1958 under the direction of Paul Bernays, influenced by Hermann Weyl. Thesis subject was model theory, which he pursued in positions at U. of Minnesota and UC Berkeley, branching out into its relations to universal algebra, infinitary logic and category theory. Back at the ETH in 1972, he worked in the foundations of computer science, first from a model-theoretic standpoint regarding semantics of programming languages, then by using combinatory logic and its models for understanding the concept of computability. Extending this to various uses in computational algebra and applied mathematics, in collaboration with his students, necessitated obtaining equipment and computing power for ETH in which he was instrumental. In the last years his work was in neuroscience on mathematical models of brain functions by the use of an interpretation in combinatory algebra.

Chapter 6
The Two Sides of Modern Axiomatics: Dedekind and Peano, Hilbert and Bourbaki

José Ferreirós

> "*No concept is univocal in the mathematical sense*",
> not even "*the number concept*". (Kronecker [32])[1]

Abstract This chapter focuses on two different facets of axiomatics: 1. the formal-logical side, linked to careful, rigorous establishing of the inferential structure of a theory, and 2. the conceptual-mathematical side, often linked to the establishment of new interconnections in mathematics, or 'deeper' ways of grounding some results. We explore this question, first by offering some classic examples in papers due to Hilbert and Bourbaki (and Hausdorff), and then going on to a simple example: the treatment of arithmetic around 1889 in the hands of Dedekind and Peano. In this thumbnail example, one can already find the above-mentioned duality. It is usual to insist on the equivalence of the works of Peano and Dedekind, but we shall argue that they had different aims—Peano focused on elementary arithmetic and its precise formulation in a new artificial language, while Dedekind aimed to systematize and "deepen the foundations" of number theory (elementary or not). We offer arguments for these claims, including a discussion of "modern" number theory in the 19th century, and we close with some philosophical remarks.

Mathematical thought is constantly evolving, "les mathématiques sont un devenir" (J. Cavaillès), so one can hardly aspire to give a general definition of this discipline. Yet an ancient idea that was eventually revitalized by the Grassmann brothers can do quite a reasonable job—the idea of a *Formenlehre*. Mathematics can be regarded as

[1] From the lecture course (Kronecker [32], 231). In context: "Will man aber z.B. den Begriff der Zahl erörtern, so muß man denselben im allerengsten Sinn, nämlich als Anzahl fassen und darf ihm nicht beimischen, was ursprünglich nicht darin liegt. Eindeutig kann man freilich den Begriff nicht fixieren, da es überhaupt keinen eindeutigen Begriff im mathematischen Sinne giebt, aber die Vieldeutigkeit muß so gering wie möglich sein.".

J. Ferreirós (✉)
Universidad de Sevilla, Seville, Spain
e-mail: josef@us.es

F. Ferreira et al. (eds.), *Axiomatic Thinking I*,
https://doi.org/10.1007/978-3-030-77657-2_6

the *science of forms*, but it's important to consider that this can be understood in two ways. One side of the idea (math as formal science) is to focus on proof, on deductive structuring of theories, on axioms, on formal processes of inference; the other side consists in the characterization and study of abstract 'forms' such as algebraic or topological structures, including of course the elaboration of new structures (at low levels, this includes the study of patterns). Let us keep both aspects in mind, avoiding one-sidedness.

I try to offer some examples of the two sides just mentioned. Let me first refer to the Bourbaki group, that many-headed modern Euclid, in their *Eléments de mathématique* Vol. 1, *Théorie des ensembles* [5]. I'll focus not so much on what they did, but on what they aimed to do.[2] They produced a careful axiomatic system for set theory, essentially equivalent to ZFC^{-}, the Zermelo-Fraenkel axioms without Axiom of Foundation, but with Choice (AC). The most relevant aspect of Vol. 1 for us is that this part of their work was done in a very *syntactic* way, in the formalistic style of the 1930s, grounded on a basic logical system that was in essence a form of first-order logic. It may be surprising that the Bourbaki, who in general were not logic enthusiasts, would publish as Vol. 1 of their treatise such a detailed syntactic treatment of the set-theoretic basis of their work. Perhaps one can see here a reflection of the atmosphere of foundational insecurity that was so typical of the 1930s.[3]

My second example is what Felix Hausdorff did in his famous axiomatization of the central notion of topological space (*Grundzüge der Mengenlehre*, 1914, Chap. 7). Notice that, even though he wrote the most important set theory textbook of the first third of the twentieth century, he decided to treat set theory intuitively and disregard the question of securing the foundations axiomatically.[4] Yet it was crucial for him to offer a general and flexible axiomatic characterization/definition of what a topological space is; he determined the new idea of a topological space, carving it out axiomatically. Following Hausdorff, a *topological space* is a set E of points x associated with subsets Ux, called *neighbourhoods*, for which the axioms hold:

(A) To each point x corresponds a Ux; each neighbourhood Ux contains point x.
(B) Given Ux, Vx two neighbourhoods of x, there is a neighbourhood Wx such that $\mathrm{W}x \subseteq \mathrm{U}x \cap \mathrm{V}x$.
(C) If point x belongs to Uy, there is a neighbourhood Ux such that $\mathrm{U}x \subseteq \mathrm{U}y$;
(D) Given two different points x, y, there are two neighbourhoods Ux, Uy such that $\mathrm{U}x \cap \mathrm{U}y = \emptyset$. [26], 213).

Hausdorff had searched among different known basic notions, trying to find a simple, flexible axiomatics, a precise but non-redundant axiom system. The Bourbaki

[2] On this topic see Corry [8], Anacona et al. [1], Mathias [32]. Let's disregard the interesting topic of how they treated AC, following Hilbert, the fact is they included the axiom of Choice.

[3] Ferreirós [16], Chap. 10.

[4] This is the question that Zermelo, like the Bourbaki later, like Whitehead & Russell, was studying at the time.

wrote that Hausdorff was able to choose the "axioms that could provide his theory with all the desirable precision and at the same time all the desirable generality."[5]

Thus, a mathematician who was relatively uninterested in securing the foundations of set theory the axiomatic way, could have a keen interest in axiomatically characterizing the abstract 'form' or structure of topological spaces. Interest in one side of modern axiomatics does not imply an equal concern for the other. Yet authors primarily interested in the second side, like the Bourbaki, could also at times be concerned with the syntactic formalism underlying a foundational theory. In sum, we have two different facets of mathematical axiomatics:

A. the **formal-logical** side, linked to careful, rigorous establishing of the inferential structure of a theory,
B. the **conceptual-mathematical** side, the elaboration of 'forms', often linked to the establishment of new interconnections in mathematics, or 'deeper' ways of grounding some results.[6]

In the following we shall further explore this question, first by offering some more classic examples and reflections, and then going to a simple example, the treatment of arithmetic around 1889, where we can already find the duality. I'd like to concentrate on this case, since it may be instructive due to its very simplicity.

6.1 Some Classic Reflections on the Two Sides

In a related vein, commenting on the evolution of Hilbert's work, Sieg [44] makes a fundamental contrast between the formal axiomatics (A.) of his late work in the 1920s, and structural axiomatics (B.) which can already be found around 1900. David Hilbert and the Bourbaki were the champions of axiomatics in that golden age, the early and mid twentieth century. By that they meant both aspects or facets mentioned above, but were they equally interested? One may say that, for both of them, it was the conceptual-mathematical aspect B. that mattered most. Of course I don't mean to deny that Hilbert was very interested in aspect A., in fact more so than the Bourbaki people. But facet B. is the one that is central for the progress of mathematical knowledge. It has to do with the identification of relevant frameworks or relational systems, the "networks of concepts" [*Fachwerke von Begriffen*]; and it is also linked with the "deepening of foundations" that Hilbert viewed as crucial to the progress of mathematics.[7]

It's therefore interesting to consider the explicit reflections that Hilbert and Bourbaki provided about axiomatics. Let us consider two famous papers, that however

---

[5] Bourbaki [6], p. 197. More generally see James [29], on Hausdorff's "classic text" in pp. 17–18, 213*ff*.

[6] We may consider this to constitute the "semantic" aspect of axiomatics, in parallel with A. the syntactic side. But many terms have several meanings in our language: with 'semantic' I would not mean formal semantics, but rather cognitive semantics.

[7] Hilbert [27], 1107–1109. This notion of deepening is explained below, in Sect. 6.1.2.

may have some defects: Hilbert's [27] 'Axiomatic Thinking' (just a century old at the time of the talk on which this paper is based) and Bourbaki's [4] 'The Architecture of Modern Mathematics'.

6.1.1 Bourbaki in 1950

In their presentation, actually due to Jean Dieudonné, the Bourbaki provide a very interesting survey of the roles of the axiomatic method in the architecture of mathematics. Besides considering the role of logical formalism and the deductive method, Bourbaki shows how a combination of analysis and synthesis leads to the core notion of *structure.* Mathematics, says Bourbaki, is a "storehouse of abstract forms," the structures—from the celebrated "mother structures" to central hybrids like ***R***, the real number system, a.k.a. the continuum. The main kinds of structure (algebraic, topological, order structures) are the *mother-structures*,[8] and it is shown how they make possible a general survey and classification of all of mathematics. Together, the axiomatic method and the structures operate a standardization of mathematical tools and methods, making possible a "remarkable economy of thought".

Throughout the paper, Bourbaki places emphasis on facet B. almost to the point of despising A.:

> It is only in this sense of the word 'form' that one can call the axiomatic method a 'formalism'. The unity which it gives to mathematics is not the armour of formal logic, the unity of a lifeless skeleton; it is the nutritive fluid of an organism at the height of its development, the supple and fertile research instrument to which all the great mathematical thinkers since Gauss have contributed, all those who, in the words of Lejeune-Dirichlet, have always labored to 'substitute ideas for calculations'. (Bourbaki [4], 1276)

The ultimate goal of axiomatics, they argue, is what the logical formalism by itself cannot provide: "the profound intelligibility of mathematics" (Bourbaki [4], 1268).

For a presentation made in 1950, this paper does not sufficiently emphasize some central mathematical notions, in particular the idea of *morphism*,[9] nor—more generally—the *interrelations* of structures that are so essential to structural math, but which at the time *had not (yet) been axiomatized*. An early, key example of such interrelations is the interplay between groups and fields that lies at the core of Galois theory. The object of this theory is to find out about the roots of equations, i.e. about the "splitting field" K which extends the "ground field" F from which the coefficients of the equation come. But in order to do so, Galois associated a group to the given equation, in modern terms the Galois group *Gal* (K/F), and the key idea is that

---

[8] The list is provisional (Bourbaki [4], 1274): "nothing is farther from the axiomatic method than a static conception of the science. ... The structures are not immutable, neither in number nor in their essential contents. It is quite possible that the future development of mathematics may increase the number of fundamental structures, revealing the fruitfulness of new axioms, or of new combinations of axioms".

[9] This word does not appear a single time, in the paper, nor even the basic notion of isomorphism. One also looks in vain for the general idea of a mapping.

subgroups of the Galois group correspond exactly to the fields lying between K and F. The Fundamental Theorem of Galois theory is in fact the detailed expression of that correspondence.[10]

Another example is the interplay between homology groups and topological spaces, that topologists in the 1920s learnt from Emmy Noether (see McLarty [35]). The examples could be multiplied at will, and one couldn't exaggerate the centrality that this phenomenon has in modern structural thinking. Hence the strangeness of finding it so absent from Dieudonné's depiction. Let me add that philosophers have reflected too little on the importance that the interrelations between different kinds of structures have in modern mathematics; this is a central aspect of recent mathematical method that is still awaiting careful treatment in structuralist philosophies.[11]

6.1.2 Hilbert in 1917

Jumping back to Hilbert's [27] paper, it's a forceful declaration of the centrality of the axiomatic method not only in math, but more generally in scientific thought. Summing up in the conclusion, Hilbert writes:

> I believe: anything at all that can be the object of scientific thought becomes dependent on the axiomatic method, and thereby indirectly on mathematics, as soon as it is ripe for the formation of a theory. By pushing ahead to ever deeper layers of axioms in the sense explained above we also win ever-deeper insights into the essence of scientific thought itself, and we become ever more conscious of the unity of our knowledge. In the sign of the axiomatic method, mathematics is summoned to a leading role in science. (Hilbert [27], 1115)

The paper has two very different facets. On the one hand, it aims to show the ample spread of the axiomatic method, and that central phenomenon of the "deepening of foundations" [*Tieferlegung der Fundamente*]; this is related to aspect B. mentioned above. Once a domain of knowledge matures to the point when a true theory can be configured, a network of concepts [*Fachwerk von Begriffen*] emerges; a summary of the whole theory can be expressed in a finite system of axioms. And this phenomenon can be observed as much in purely mathematical knowledge-domains, as it is found in physics and other sciences (Hilbert [27], 1108). For this reason, the paper contains as many examples from the field of physics as from within math.

On the other hand, Hilbert is interested in listing a series of questions that he aims to attack from a new perspective. These questions point in the direction of the *proof theory* or *Beweisstheorie* that would become famous in the 1920s. The first facet of the paper makes it seem backward-looking, linked very much with Hilbert's early work and with his interest in physics (but not quite related with the Hilbert Program), while this second side has the opposite effect, and points toward facet A.

---

[10] See e.g. http://mathworld.wolfram.com/FundamentalTheoremofGaloisTheory.html (accessed 4.03.2019); concerning the history of Galois theory, see Neumann [36], Ehrhardt [13].

[11] At least in the usual ones, e.g., the versions of structuralism due to Shapiro, etc. The situation is different with category-theoretic structuralism, since category theory is precisely an explicit way of capturing those interrelations.

Hilbert believes that physics and epistemology (*Erkenntnistheorie*) are the disciplines that lie closer to math, and the new questions I have just mentioned are in fact described as "erkenntnistheoretischer," epistemological: the problem of the solvability of every mathematical question; the controllability of the results of research; finding a criterion of simplicity for proofs; the relation between content (*Inhaltlichkeit*) and *Formalismus* in mathematics and logic; and, last but not least, the problem of decidability (*Entscheidbarkeit*) of a given question by finitely many operations.

All such "questions of principle," and a few more discussed before the passage just mentioned (independence and consistency),[12] establish the need to study the "essence of mathematical proof in itself" (Hilbert [27], 1115). There is, for instance, a need to show the absolute consistency of arithmetic and set theory, for math is a paradigm of rigorous science, and this idea must be vindicated; that will require "the axiomatization of logic" too. The very idea of proof must be thematized and made objective, becoming "itself an object of investigation, just as the astronomer considers the movement of his position, the physicist studies the theory of his apparatus, and the philosopher criticizes reason itself" (Hilbert [27], 1115). Proof theory will form an important new field of research which remains to be developed.

As I mentioned, this paper has the peculiarity that Hilbert gives many examples from physics. Many sentences are devoted to statics, mechanics, thermodynamics, gas theory, radiation theory, electrodynamics; and in fact initially the theories described are quite far from what we would call *axiomatized* in the strict sense.[13] For instance:

> Even for successful theories that have long been accepted, it is difficult to know that they are internally consistent: I remind you of the reversibility and recurrence paradox in the kinetic theory of gases. ... particularly in physics it is not sufficient that the propositions of a theory be in harmony with each other; there remains the requirement that they not contradict the propositions of a neighbouring field of knowledge. (Hilbert [27], 1111)

Even among the mathematical cases discussed at the beginning, Hilbert's choice is rather surprising: the theory of surfaces, the theory of primes (based on Riemann's ζ-function), or the theory of equations (based on the postulate of the existence of roots):

> If in establishing the theory of Galois equations we assume as an axiom the existence of roots of an equation, then this is certainly a dependent axiom; for, as Gauss was the first to show, that existence theorem can be proved from the axioms of arithmetic. (Hilbert [27], 1110)

[12] To be more complete, the latter are: 1. Dependence or independence of the sentences of a theory, in particular of the axioms; 2. the role of the axiom of continuity in a given theory;* 3. freedom-from-contradiction of a theory, in a relative or an absolute sense; and 4. the *Entscheidungsproblem*. Concerning *, he writes: (Hilbert [27], 1110): "For example, if one follows Planck and derives the second law of thermodynamics from the axiom of the impossibility of a perpetuum mobile of the second sort, then this axiom of continuity must be used in the derivation".

[13] On this general topic, see L. Corry [9].

What is meant here is the fundamental theorem of algebra, proved by Gauss in his 1799 dissertation and first proved by him rigorously in 1816. Hilbert suggests a process of deepening the foundations that may start with this central mathematical result, and push ahead to "ever deeper layers" of axioms, e.g., axioms for the field of complex numbers, or even more general axioms of field theory, or the basic axioms of set theory.[14] Thus "we also win ever-deeper insights into the essence of scientific thought itself" (cited above).

Galois Theory is mentioned several times by Hilbert as a good and central example of axiomatics (see Hilbert [27], 1112–1113), but in fact to our eyes the situation is interesting or even surprising. To begin with, it's actually difficult to figure out what Hilbert means by "Galois Theory" at this point in time; we cannot enter into this topic, for it would lead us too far afield into a historical discussion. It's likely that he had in mind at most the axiomatizations of general field theory and group theory (and their links with basic set theory), but it seems also probable that he was not reflecting on the Fundamental Theorem of Galois theory (the precise interrelations between a series of field structures and a series of groups) and how to ground it.[15] In 1917, to the best of my knowledge, Galois theory the way we know it existed but, precisely, such absolutely central aspects of the theory had not been axiomatized. A satisfactory axiomatic treatment of Galois theory, in a modern structural way, must tackle the *interrelation* of field and group structures, later presented as a matter of functors from one category (fields) to another (groups).

Morphisms and functors play a key role in structural thinking: they allow mathematicians to ignore extra structure, or to add supplementary structure, according to the situation at hand, so that mathematical obstacles can be superseded. It's interesting to remark that this facet of modern mathematics had been presented long time before by Dedekind, who underscored the role of mappings and morphisms, and who prepared some key aspects of the modern way of presenting Galois theory (see Ferreirós [20], Ferreirós and Reck [21]).

6.2 Grounding, Bridging, and Deepening

The 'semantic' aspect of axiomatic thinking, what I called the conceptual-mathematical facet, requires to *search for new concepts and new frameworks, in a creative way*. This is quite unlike A. the formal-logical side, at least in cases when the theory to be formalized is known, and the logical framework is regarded as given (e.g. first-order logic). Here is yet another way of looking at the two facets: the formal-logical is related with the *grounding* of important theories, with the foundational

[14] Hilbert had a definitely set-theoretic perspective on math: the ZFC axioms ground both the number systems (in particular the complex numbers) and the general theory of fields. In 1910 he had described set theory as "that mathematical discipline which today occupies an outstanding role in our science, and radiates its powerful influence into all branches of mathematics" (cited in Ferreirós [17]).

[15] A referee suggests that he may have had in mind the paper Weber [49].

securing e.g. of core structures such as $\boldsymbol{N}$ and $\boldsymbol{R}$ and $\boldsymbol{V}$,[16] and so on (paradigmatic work of this kind was done by Dedekind, Peano, Hilbert, Zermelo, Tarski). Meanwhile, the conceptual-mathematical is related to innovation and to *bridging*, to the establishment of new ground levels (e.g. set-threoretic notions in the late nineteenth century; or arrows & objects & functors in category theory) or the elaboration of interconnections between different domains of mathematics. Sometimes it has to do with the isolation of the right *conceptual tools* for creating bridges (morphisms are a key example).

The formal logical or *grounding* aspect is obviously linked with the logical and inferential side of things, rigorous proof and exhaustive analysis—with "logic" as theoretical hygiene, as Poincaré or Weyl said. The conceptual-mathematical aspect is much more connected with the rethinking of results, with new "intuitions" or deeper insights. Intuition being a rather obscure word, too equivocal as it has been employed in too many different senses, it's employed here to point towards a phenomenon that would need to be explained or analysed.

You may ask, what is the nature of the relation between aspects A. and B., and the ideas of grounding and bridging?[17] If you look at it from a purely logical viewpoint, these connections may not be clear at all. But introducing the historical dimension, the development of mathematical ideas, things should be clearer. A formal axiomatic reconstruction of some part of mathematics first appears (at least in many relevant cases, central examples like Euclid's geometry, Peano's arithmetic, Dedekind's theory of real numbers, or Zermelo's set theory) when there is previously a well-developed body of theoretical results, and some mathematician aims to systematically present this body, grounding it in some explicit principles. Hilbert [27] is devoted to a reflection on this historical fact, at least one half of the paper.[18] Choices have to be made about the basic notions to be employed, and the particular selection of axioms, which of course doesn't have a unique answer—different options for *grounding* are available.

As for the link between B. and *bridging*, the example of Galois theory should be exemplary. Clarifying the question of the solvability of equations by radicals, as it was done around 1830, required the introduction of new concepts that were far from the initial concentration on formulas and equations (which had almost defined math in previous times).[19] Instead of looking for new ingenious manipulations of the formula or its components—the equation in question, say a fifth-degree one–, the solution was found by introducing a new concept of group, specifically defining the Galois group of a given equation, and developing tools for the analysis of its properties (solvability, normal subgroups). Let me underscore that the way the idea of group appeared with (Lagrange and) Galois was a great surprise, as nobody expected that

[16] $\boldsymbol{V}$ is conventionally employed to denote "the" intended model of axiomatic set theory, the universe of sets—although some set-theorists actually endorse a multi-verse perspective. See Arrigoni [2].

[17] I thank an unknown referee for insisting that I should consider this question more carefully.

[18] The other half introduces new questions such as the problems of consistency or decidability, see above.

[19] For details, see e.g. Kleiner [30], or Wussing [50].

the analysis of roots of equations would have to be approached with such notions. The concept of a group was new, the idea that one can associate with each equation 'its' group was a complete novelty, as was the notion that one can 'read' properties of the equation from consideration of the properties of its Galois group.

A surprising bridge had been established between a traditional topic and a series of new mathematical ideas. Thus we find a clear case of bridging in connection with the introduction of new concepts, with conceptual and semantic innovation. (In due course the same pathway led to the concept of a number field, which added clarity to the whole issue.) The case of Galois theory is by no means unique. A similar basic example is given already by Cartesian geometry, bridging between the area of synthetic geometry and the algebraic study of equations; likewise, among very basic topics, for the geometric interpretation of the complex numbers. And bridging with conceptual innovations is a central aspect of the modern structuralist methodology—one that unfortunately has not sufficiently been reflected upon by philosophers.

Another relevant historical example is the elaboration of the notion of manifold in connection with differential geometry and topology[20]; the notion of a manifold was very abstract already with Riemann, constituting a relevant break with past conceptions of mathematics. Or, to continue with the case of groups, their emergence in the context of Klein's Erlangen Program, shedding new light on the links between different types of geometries; but also the discovery of their role in studying crucial aspects of quantum theory, e.g. related to spectral lines.

Thus, grounding is constitutive of A. the formal-logical aspect, and there seems to be no crucial instance of bridging without B. the conceptual-mathematical aspect entering into the picture. This doesn't mean that there cannot be instances of grounding that also feature conceptual innovation; we'll see an example in Sect. 6.3. Indeed, deeper grounding (the *Tieferlegung* of foundations so dear to Hilbert) is very often, if not always, a case in point. Consider e.g. the way in which axioms for the real numbers can be grounded in set theory, which was actually one of the relevant historical pathways by which axiomatic set theory was elaborated (work of Dedekind, Cantor, Hilbert). Here perhaps, to make my claim more clear, I should make explicit my view that the idea of set was a conceptual innovation introduced after 1850 (see the Epilogue to Ferreirós [16]).[21]

6.3 On the Axiom Systems of Dedekind and Peano

I have promised to offer a simple example where we can already find the duality A./B., namely the treatment of arithmetic around 1889. My exposition of this case will be

[20] For details, see Scholz [42], or Ferreirós [16], Chap. 2.

[21] To make a key point explicit, briefly, let me use Russell's terminology: a set is a class-as-one, whenever a is a set one can write $a \varepsilon x$; but the classes that we find in the work of logicians prior to 1850 are either conceived intensionally, or they can be understood as a class-as-many.

very close to the actual history, but not completely; I will perhaps overemphasize one aspect of Peano's work. Normally people insist on the equivalence of the works of Peano and Dedekind on elementary arithmetic, the basic theory of the structure of the natural numbers. The different axiomatizations or systematizations they produced are seen as conceptual variants. Yet Peano himself saw them as different.[22]

Peano's treatment is more A.-style: grounding of a well-known theory, precise formulation in a new artificial language, exhaustive analysis based on Grassmann's previous work [25]. Quite important in this analysis is that Peano formulated explicitly as a new axiom the *principle of induction*, which Grassmann had left implicit—this is quite surprising and a clear defect, because his approach relied on it heavily![23] The Induction principle says that any property that applies to 1 and is *hereditary* (i.e., transfers from a to $a + 1$) is a property of all numbers, but also that any set containing 1 and closed for the step $a \rightarrow a + 1$, must contain $\boldsymbol{N}$. We may also interpret the Induction axiom as saying that the set $\boldsymbol{N}$ is minimal—and that guarantees complete induction.

I will argue that, by contrast, Dedekind's treatment is B.-style; but let us first have a look at Peano's work.

3.1. It will be worthwhile to compare the original formulation of Peano's axioms (1889) and his second paper on the topic (1891). To facilitate reading, I modernize Peano's logical notation slightly (compare [37], 94)[24]:

1. $1 \; \varepsilon \; \mathrm{N}$.
6. $\mathrm{a} \; \varepsilon \; \mathrm{N} \supset \mathrm{a} + 1 \; \varepsilon \; \mathrm{N}$.
7. $\mathrm{a}, \mathrm{b} \; \varepsilon \; \mathrm{N} \supset (\mathrm{a} = \mathrm{b} \leftrightarrow \mathrm{a} + 1 = \mathrm{b} + 1)$.
8. $\mathrm{a} \; \varepsilon \; \mathrm{N} \supset - (\mathrm{a} + 1 = 1)$.
9. $[\mathrm{k} \; \varepsilon \; \mathrm{K} \wedge 1 \; \varepsilon \; \mathrm{k} \wedge \mathrm{x} \; \varepsilon \; \mathrm{N} \wedge (\mathrm{x} \; \varepsilon \; \mathrm{k} \supset_{\mathrm{x}} \mathrm{x} + 1 \; \varepsilon \; \mathrm{k})] \supset (\mathrm{N} \supset \mathrm{k})$.

The last instance of '$\supset$' in 9. denotes set inclusion (in modern symbols, $\mathrm{N} \subset \mathrm{k}$), all the previous ones are mere conditionals. '$1 \; \varepsilon \; \mathrm{N}$' was read, alternatively, '1 is a number' and '1 belongs to (the class) N'. Peano's symbolism can always be read dually, in terms of properties and in terms of sets. Axiom 9. can be read thus: if k is a class, $1 \; \varepsilon \; \mathrm{k}$ and, for $\mathrm{x} \; \varepsilon \; \mathrm{N}$, whenever $\mathrm{x} \; \varepsilon \; \mathrm{k}$ then also $\mathrm{x} + 1 \; \varepsilon \; \mathrm{k}$, then the class N is included in k.

[22] I mentioned the differences between Dedekind and Peano already in Ferreirós ([15], 622), but my indications there—important in my view—were probably too quickly explained. For further details see also Gillies [23], Segre [42], my paper, or Skof [43].

[23] Even if Grassmann's intention was not to provide an axiomatic foundation of arithmetic (probably because he had in mind a deeper grounding), he should have clearly stated the general principle of induction. See Grassmann [25], 6–7 where the first "inductorische" proof can be found. His textbook aimed to present arithmetic "in its most rigorous form" (Preface, vi).

[24] I replace some dots by parentheses, employ the symbol '$\wedge$' for conjunction ('and'), and replace an occurrence of '=' in 7., which denotes 'if and only if', by '$\leftrightarrow$'. This first version also intermingled the principles of the logic of identity (restricted to number objects—axioms 2–5) with the famous five axioms (Peano [37], 94).

In the second version, we have five axioms (Peano [38], 90), some of which we will need to gloss (at the right I indicate the corresponding condition in Dedekind's definition of 'simply infinite set', [10], point 71):

1. $1 \varepsilon \mathrm{N}$.
2. $+ \varepsilon \mathrm{N} \mid \mathrm{N}$. condition α [$\mathrm{N}+ \subset \mathrm{N}$]
3. $a, \mathrm{b} \varepsilon \mathrm{N}\ (\mathrm{a}+ = \mathrm{b}+ \ \supset \mathrm{a} = \mathrm{b})$. condition δ [similarity]
4. $-\ (1 \varepsilon \mathrm{N}+)$. condition γ
5. $[\mathrm{s} \varepsilon \mathrm{K} \wedge 1 \varepsilon \mathrm{s} \wedge (\mathrm{s}+ \ \supset \mathrm{s})] \supset (\mathrm{N} \supset \mathrm{k})$. condition β [chain axiom]

'*a*+' now indicates the successor of *a*, with + understood to be a function; more standardly one might write $+(a)$. Axiom 2. states that this is a function from N to N, since 'N | N' denotes the collection or class of such functions. The change in the fourth axiom may seem subtle, but it states that 1 is not an element of the image-set N+; this 'set-theoretic' formulation is different from the comparable axiom in Peano [37].

One can easily notice that the axiom system has been reconceived in terms of the language of functions (or mappings) and images. Peano himself underscores the fact that the novelty in this paper consists in resting on the general notion of function ([37], 87) and thus making more general and concise the treatment of arithmetic. But even the induction axiom is significantly different, despite appearances: 's+ ⊃ s' means that the image-set s+ is included in s.[25]

If one reads Peano [38] more carefully (p. 93), it becomes apparent that his intention was not to present his own axiom system of two years earlier, but to present Dedekind's system (see also pp. 89, 90, 94). Peano seems to have perceived some key differences between his approach to arithmetic and Dedekind's: when presenting Dedekind's approach, he formulates the axioms as conditions on a set ***N***, its subsets, and a successor *function* (mapping). When the axioms are his own (1889 and later works) they are formulated more elementarily as conditions on the *elements* of a set *and* on the basic arithmetic *operation* of adding 1 (see Ferreirós [15] 622). In particular, Peano in (1889) understood 'a + 1' as a new term, formed combining three symbols; but two years later *a*+ is the image of *a*, the result of application of a function or mapping.

It seems that Peano reflected carefully on this difference between his elementary analysis and the more abstract one of Dedekind: the latter avoids to focus on elements and basic operations, to prefer a 'set-theoretic' and map-theoretic approach underscoring the set ***N***, its subsets, and maps. One avoids, if possible, conditions expressed in terms of elements and operations on elements, to prefer explicitly set-theoretic or map-theoretic conditions. This agrees quite well with the meaning that Emmy Noether would give to the expression "set-theoretic" in the 1920s.[26] Because Peano's

[25] In the usual modern way, 'a+' would be written '+(a)', and 's+' written '+(S)'.

[26] See McLarty [35]. As Ø. Ore wrote in 1935, Noether's tendency in the discussion of the structure of algebraic domains was this: "one is not primarily interested in the *elements* of these domains but in the relations of certain *distinguished subdomains*" (cited in McLarty [35], 194). Also W. Krull underscore another Noether's principle: "base all of algebra so far as possible on consideration of

approach is more elementary, his system is easy to render in first-order logic,[27] while Dedekind's system is very explicitly higher-order or set-theoretic.

It's interesting to notice that this was not the only time that Peano presented Dedekind's work carefully. In fact, he alternated between publications that present his elementary axiomatics, and other publications with Dedekind's set- and map-theoretic axioms. Most importantly, in the 1898 edition of the *Formulario mathematico* he even produced a summary of Dedekind [10], based on extracts of this work presented *verbatim*. It's clear that Peano admired Dedekind's contribution and was interested in facilitating its wide diffusion.

3.2. What Dedekind provided was not a mere systematization and rigorous development, but a "deepening of the foundations" of basic arithmetic. He was rethinking results, and in the process obtaining new and deeper insights, based on a *new conceptual basis*—with *set* and *map* as the new basic notions. Employing the idea of set to ground arithmetic was standard, indeed so proceeded Schröder in the early 1870s, Peano in 1889, as well as Cantor and several others. But the general notion of a map, as a generalization of function, had never before been thematized. Moreover Dedekind established a foundation for the number concept starting from infinite sets, in a reversal of the traditional way of thinking. To do so, infinity had to be defined independently of numeration, which is what the definition of Dedekind-infinite does[28] (a Dedekind-infinite set comes endowed with a(n injective) map s: S $\rightarrow$ S, and in the particular case of $\boldsymbol{N}$ this provides a well-ordering of the set).

Next, in order to reduce the crucial principle of induction to that conceptual basis, Dedekind found it necessary to elaborate a sophisticated theory of "chains" (see [10], Sect. 4, [15]). This is not the place to enter into the subject, so it will suffice to say that the "chain condition" $\boldsymbol{N} = s_o\{1\}$ (Dedekind's condition β, read: $\boldsymbol{N}$ is the s-chain of singleton $\{1\}$,[29] s being a map) guarantees minimality of the natural-number structure. The number structure $\boldsymbol{N}$ is the minimal set which includes $\{1\}$ and is closed under the successor function s: $\boldsymbol{N} \rightarrow \boldsymbol{N}$, and thus Induction will hold. In fact, the context of chain theory allowed Dedekind to significantly generalize the principle of Induction.

But Peano, in his rendering of Dedekind's axioms, decided to sidestep this complication (Peano [38], 93). He produced a modified version of his own Induction axiom, only slightly closer to Dedekind's preference for 'set-theoretic' formulations (s+ $\subset$ s), avoiding conditions expressed in terms of the elements and basic relations on them (x ε k $\rightarrow$ x + 1 ε k, for all x). The difference does not spring to the eye when considering the symbolic rendering of the axioms, but a careful consideration of the systems makes it clear. Peano showed his skill and elegance in handling symbolic

isomorphisms" (op. cit. 194). Both ingredients can be found in Dedekind already—thus I'd qualify what McLarty says in [35], 193 (see also Ferreirós [20]).

[27] Except for the well-known problem of the Induction principle, which must be turned into an axiom-schema, as T. Skolem suggested for the first time.

[28] A set S is *Dedekind-infinite* if and only if there is a proper subset T of S and a one-to-one map g: T $\rightarrow$ S (i.e., the subset T is bijectable with all of S).

[29] Here I modernize and improve his notation, in fact Dedekind wrote '$s_o(1)$' or '1_o', exploiting a (dangerous) ambiguity between 1 and $\{1\}$. He explained the danger himself in some manuscripts.

calculi, by producing (in Peano [38]) a kind of mixture of his previous system and Dedekind's.

Coming back to Dedekind's chain theory, it was very general and proved to be of great relevance to general set theory. Chain theory allowed Dedekind to prove the Cantor-Bernstein theorem, a central result in the theory of transfinite cardinalities which Cantor had formulated but found himself unable to prove (see Hinkis [28], Sieg [44]). Zermelo later, in 1908, produced a transfinite extension of the theory of chains that was key to his second, definitive proof of the Well-Ordering theorem [51]. Skolem [46] employed chain theory in his work on the Löwenheim-Skolem theorem.

Also relevant to mention is the fact that Dedekind established a foundation for recursive definitions, which at the same time became the basis for other general results (isomorphism of models of his axioms, and the establishment of the notion of cardinal number). Meanwhile, Peano employed recursive definitions of sum and product, like Grassmann before him, but they were introduced intuitively and in effect contradicted Peano's explicit doctrine on definitions.

3.3. But there is even more. If one asks, what is the ultimate source of the differences between the work of Peano and Dedekind, the answer seems to be that, in a sense, they were not aiming at a systematization of the same theory. In the introduction to *Arithmetices principia*, Peano wrote: "Questions that pertain to the foundations of mathematics … still lack a satisfactory solution. The difficulty has its main source in the ambiguities of language" (Peano [37], 85). His logico-foundational work was not aiming at deepening the foundations, but had a more direct and pragmatic goal: to develop an artificial, symbolic language well-suited for mathematics, and to employ it in order to present mathematics in a perfectly precise, unambiguous way. The treatment of basic arithmetic was just a key instance of the application of such methods.

By contrast, Dedekind's objective was broader. This is consistent with the fact that his foundational researches were always linked with reflections about advanced mathematical work, aiming to *unify the methods* and the theoretical tools employed in both areas: the foundations and the extensions (see Ferreirós [16], Chap. 7). To present the situation most directly, one can say: Dedekind's aim was not just to axiomatize elementary arithmetic, but rather to axiomatize (advanced) *number theory*.

Isn't this the same thing? Not really. This contrast can be found already in Gauss's *Disquisitiones arithmeticae*, where he distinguishes *elementary* arithmetic (dealing with the basic properties of numbers, which apply in general to all of them) and *higher* arithmetic (*Arithmetica sublimior*), dealing with research into the properties of particular kinds of numbers, e.g. the primes[30] (Gauss [22], *praefatio*). It's the second that we call number theory. But since the time of Gauss's *Disquisitiones*, number theory had grown immensely, both in the direction of *analytical* number theory (the application of methods of analysis to prove number-theoretic results) and *algebraic* number theory (the expansion of number-theoretic questions and methods

[30] An example is the law of asymptotic distribution of prime numbers among the naturals, the Prime Number theorem proved in 1896 by Hadamard and de la Vallée Poussin (see below).

to new domains, in particular to integers in algebraic number fields).[31] My claim is that Dedekind's aim was to systematize and found *number theory* in this very general sense of the expression.

Let's not create any confusion: the material developed and proved *in detail*, theorem after theorem, in Dedekind [10] does not go beyond the natural numbers, in the sense of Gauss's elementary arithmetic, *but* the prefaces make it plainly clear that the goal is much more ambitious—and, most importantly, this squares with every detail about the methods and basic tools employed. Dedekind is explicit saying that his theory suffices for developing the entire number system, all the way up to the real $\boldsymbol{R}$ and the complex numbers $\boldsymbol{C}$, together with their algebra and analysis.[32] The important aspect, for us here, is that he sees set theory and map theory as the general basis for a reduction of *arithmetic, algebra and analysis*, which makes it possible to systematize all of the number theory of his time. The same foundation established for basic arithmetic in this work suffices to develop analytical number theory and algebraic number theory. Thus the theory covered by Dedekind's [10] deeper foundation is much wider than the elementary theory which constituted the target for Peano [37].[33]

6.4 Modern Number Theory in the Late Nineteenth Century

In the interest of making clearer the difference between the two theories—what I am calling elementary arithmetic, and what I call (advanced) number theory—I shall now provide some more information about the historical background of my reconstruction. Thus I'll say a few words about the extensions of number theory in the nineteenth century; yet readers may choose to jump to the final section of this paper.

In the wake of some contributions of Gauss and Dirichlet during the 1830s, the field of number theory was expanded in several directions. Next to Gauss and Dedekind, Gustav Lejeune-Dirichlet was one of the great masters of number theory. In 1837 he published a very noteworthy theorem on the presence of infinitely many primes in some arithmetic progressions; the title is clear and reads: "every unbounded arithmetic progression, whose first term and common difference are integers without

---

[31] This means that the topic is no longer the natural numbers, but the integers in any number-field $K \subset \mathbb{C}$, a simple example being the Gaussian integers $a + ib$ (where i is the complex unit).

[32] He actually says: "the other extensions may easily be carried out in the same way," and he reserves himself the right [*ich behalte mir vor*] "sometime to present this whole subject in systematic form" (Dedekind [10], 792).

[33] The key idea is implicit in Dedekind's well known thesis that "arithmetic (algebra, analysis) is only a part of logic"—in a word, *all of pure mathematics* stands on the same foundation, whose basis is laid in Dedekind [10]. This is not the place to discuss Dedekind's notion of logic, nor his form of logicism and its place in history (see Benis Sinaceur [3], Ferreirós [18], Reck [31], Klev [40]).

common factors, contains infinitely many prime numbers" [11]. This result created a stir because of its proof method: it was established by considering certain infinite series, and required principles from infinitesimal analysis. It was the birth of analytical number theory, which would later lead to Riemann's work, including the Riemann Conjecture, and to the Prime Number Theorem, which establishes the law of distribution of primes i.e. their frequency (first conjectured by Legendre and Gauss, proved by Hadamard and De la Vallée Poussin in 1896, independently, using methods of complex analysis). Dirichlet himself emphasized the "novelty of the principles employed" in his "completely rigorous proof", acknowledging that it "is not purely arithmetic, but rests in part on the consideration of continuously varying magnitudes" (Dirichlet [11], 316).[34]

In the subsequent debate, mathematicians came to conjecture that the theorem, albeit easy to state in simple arithmetic terms, might be impossible to prove without analytical means. As late as 1921 an expert like G. H. Hardy was still emphasizing that "no elementary proof is known" and conjecturing that it is "extraordinarily unlikely" that one would be found.[35] Already in the mid-nineteenth century, this state of affairs prompted foundational and methodological reflections. What have the properties of prime numbers to do with infinite series? How to explain the need for infinitary methods in pure number theory? And, does this tell us something about the nature of advanced algebra and analysis? The story we are telling was in fact intimately related with the emergence of the 'arithmetizing' viewpoint, that algebra and analysis are just highly developed arithmetic.

Dirichlet, who was a philosophical mathematician, was in the habit of stating (according to Dedekind, who was very close to him) that every theorem of algebra and higher analysis can be expressed as a theorem about natural numbers:

> From just this point of view it appears as something self-evident and not new that every theorem of algebra and higher analysis, no matter how remote, can be expressed as a theorem about natural numbers—a declaration I have heard repeatedly from the lips of Dirichlet (Dedekind [10], 792).

The *arithmetization* of pure mathematics, which started already with Gauss, thus took a more precise and influential shape, and would subsequently be developed (in different ways) by Dedekind, Weierstrass, Kronecker, Cantor and others.[36]

The other new discipline of number theory was *algebraic* number theory—the theory of integers in fields of complex numbers. This got started with Gauss's expansion of number-theoretic research to the complex integers ($a + bi$ with a, b in $\mathbf{Z}$), in order to prove the so-called law of biquadratic reciprocity; this happened in 1832. Gauss himself conjectured that the proof of other "higher" reciprocity laws would require the introduction of new kinds of complex integers, and the study of their

---

[34] "Der von mir gefundene Beweis ... ist nicht rein arithmetisch, sondern beruht zum Theil auf der Betrachtung stetig veränderlicher Grössen".

[35] Hardy's 1921 Lecture to the Mathematical Society of Copenhagen, quoted in Chudnovsky et al. [7], p. 181. Hardy was later proved wrong by the work of A. Selberg.

[36] On the varieties of arithmetization, see Petri and Schappacher [39].

number theory; Eisenstein in fact introduced a new kind of integers, for cubic reciprocity. In later years, Ernst Kummer became the great expert in the field, who studied (put in modern terms) rings of cyclotomic integers and discovered the problem that they were not unique factorization domains. His highly creative solution to the obstacle thus found, was to introduce "ideal numbers" in order to "complement and simplify" the theory of the complex numbers he was studying (Kummer [33] vol. 1, 203).[37] He wrote about his belief "that these factors render visible, so to speak, the internal constitution of numbers, so that their essential properties are brought to light" even if they are not effective elements of the number-domain under consideration.[38]

It is well known that this became Dedekind's specialty, in which he contributed his celebrated "Ideal theory", a modern re-elaboration of Kummer's idea. Starting from his work, Dedekind and Kronecker turned the study of the new types of integers into a fully general theory, algebraic number theory; Dedekind operated with the notion of "ideals" while Kronecker preferred to work algorithmically with his "divisors" (Stillwell [47], Ferreirós [16, Chap. 3]. Dedekind's approach, one can say, reinvented the discipline of advanced number theory by introducing at all levels a new proto-structuralist, set-theoretic viewpoint (an ideal is an infinite set of integers, complying with two simple structural conditions). Ideals were operated upon as if they were common numbers, and they became the focus of the new theory.[39]

A celebrated textbook of nineteenth century number theory was the *Lectures on Number Theory,* the *Vorlesungen* due to Dirichlet and Dedekind, largely based on the former's lectures, but written by the latter.[40] The book, first published in 1863, went to several new editions in 1871, 1879 and 1894, each with a new version of ideal theory. And in fact the perspective on number theory offered by this influential work squares with what I have described. Dedekind made the effort to include appendices, *Supplemente* which presented relevant methods and results of analytical number theory and algebraic number theory. Already in the 1863 edition, Supplement VI offered a careful presentation of Dirichlet's theorem on primes in arithmetic progressions, based on Dirichlet L-series. The second edition, in 1871, featured the famous Supplement X which not only presented Gauss's theory of the composition of binary quadratic forms, but also Dedekind's own ideal theory. Indeed, to justify the decision to include this material, he asserted that ideal theory "throws a new light" on the "main subject" of the whole book "from a higher standpoint" (Dirichlet and Dedekind [12], preface). This is yet another statement of his outlook on number theory.

---

[37] "Es ist mir gelungen, die Theorie derjenigen complexen Zahlen [...] zu vervollständigen und zu vereinfachen; und zwar durch Einführung einer eigenthümlichen Art imaginarer Divisoren, welche ich ideale complexe Zahlen nenne;"

[38] *Mémoire sur la théorie des nombres complexes composés de racines de l'unité et de nombres entiers,* 1851, in Kummer [33]. Kummer was sympathetic to Hegel's ideas, and the choice of name "ideale Zahlen" may be an implicit reference to German idealism.

[39] All of this had an impact on Cantor, who started his career in Berlin as a number theorist. See Ferreirós [16].

[40] See Goldstein et al. [24].

I hope from the foregoing it will be clear to the reader that, around 1890, there was no reason to believe that a systematic basis for elementary arithmetic could be sufficient also for the derivation of key theorems of advanced number theory. In particular, the theorems of analytical number theory mentioned before were believed to transcend elementary arithmetic,[41] and the same applies to Dedekind's ideal theory, whose foundations were close to Galois theory. Thus, when I say that Peano's axioms aimed to axiomatize elementary arithmetic, while Dedekind's system had the aim to systematize advanced number theory, it should be clear that there is a relevant difference in the target theories. If I'm right, one can say that the problem tackled by Dedekind was significantly more complex and difficult.[42]

The well-known letter to Keferstein, Febr. 1890, highlights in a different way the ambitious nature of Dedekind's enterprise. He asked the question: "What are the mutually independent fundamental properties of the sequence N, i.e., properties which are not derivable from one another but from which all others follow?" (van Heijenoort [48], 99–100). This question was perfectly common to both authors. But then he adds: "How should we divest these properties from their specifically arithmetic character so that they are subsumed under more general notions, under activities of the understanding *without* which no thinking is possible at all—but *with* which a foundation is provided for the reliability and completeness of proofs, and for the construction of consistent notions and definitions?" This poses the question of deepening the foundations in a most radical way.

6.5 Concluding Remarks

I have expanded on this topic in order to make comprehensible my interpretation, that the aim of Dedekind's analysis and foundation of the natural number system was different, and in fact broader, than Peano's. The great Italian mathematician had more pragmatic aims: to perfect the language of mathematics, introducing a new symbolic notation; to analyze and bring to the fore the logical principles that underlie mathematical thought, highlighting the role of Induction. Dedekind by contrast was immersed in an ambitious project to reconstruct the mathematics of his time, from ground to top, rethinking the basic concepts and methods, delineating anew the different theoretical edifices. This affected as much the most basic fields, like arithmetic, and the most advanced, as seen in the cases of ideal theory and Galois theory [20, 21].

[41] Many years later, it turned out that Dirichlet's theorem on primes in arithmetic progressions could be proved elementarily; this was done by A. Selberg. But the problem whether a given theorem in advanced number theory allows for an elementary derivation is often a very difficult open question, even today. Best known is the case of Fermat's last theorem.

[42] In fact, Peano's *Arithmetices principia* went beyond the theory of natural numbers, including sections on the integers, the rationals, the real numbers. But the above statement remains true.

As a consequence, Peano was satisfied with a piecewise reconstruction of the basis of different mathematical theories: basic arithmetic, real-number arithmetic, geometry, and so on; focusing above all on the adequacy of his new symbolic language for such a project. He was very successful in that mathematical logic and its formalisms would keep growing in importance. Dedekind had the project of a unified treatment of pure mathematics, i.e. "all of arithmetic" meant to include advanced algebra and analysis, hence also advanced number theory. He was successful insofar as his proposals had significant impact on twentieth century set-theoretic mathematics, and above all on the twentieth-century structural style.

More generally, we have seen that axiomatics has two different sides that contribute to the evolution and perfection of mathematical concept systems and their ideography: the formal-logical side A. and the conceptual-mathematical side B. We've revised some of the reflections offered about them by key contributors to modern axiomatics, Hilbert and Bourbaki, seeing that the two sides of modern axiomatics are present in their work. The Bourbaki were mostly interested in the conceptual-mathematical aspect, and this was also the starting point for Hilbert, and a constant interest throughout his career, the 1917 paper being a clear example. Yet Hilbert's later work in the 1920s gave pride of place to the formal-logical aspect, for reasons having to do with his project to *mathematize* some epistemological questions about mathematics. This gave rise to the metamathematics of proof theory, making necessary the concentration on a formalistic approach as key method for investigating questions of consistency and decidability.

Let me close with some philosophical remarks. Can we say that the concept of number systematized by Dedekind is different from Peano's? In a sense, I believe the answer is yes, although I don't mean to suggest that both concepts are incompatible.[43] The idea would be rather that one concept extends the other, and in the process new connotations are introduced, and new possibilities for future development.

In previous work (see Ferreirós [19], 101–104) I have suggested that, as mathematics progresses, there may be conceptual diversification due to changing networks of practice & knowledge. That is to say, adding a new layer to pre-existing practices, in the form of new symbolic and theoretical frameworks, will in general have the effect of modifying or modulating the notions affected by the new systematic links. New inter-practice connections effect an alteration in the mathematician's conceptual understanding. That would be happening, in our case, with the addition of concepts and methods of algebraic and analytical number theory, but also with the systematic addition of map-theoretic thinking to elementary arithmetic. It would be naïve to interpret what happened with the introduction of a set-theoretic & map-theoretic, structural analysis of $\boldsymbol{N} = \langle \mathrm{N}, s, 0 \rangle$, as an instance of conceptual analysis that merely clarified an already-given concept. The number concept is modulated or subtly modified when the web of frameworks in which it is embedded becomes more complex.

[43] Peano was well aware of current work on number theory, and this applies both to analytical number theory and to Dedekind's ideal theory, which he had followed closely.

If conceptual understanding depends on the web of practices and knowledge that an agent commands and links with the concept in question, agents who command different webs of practice and knowledge may differ in their conceptions. This, I believe, suggests an interesting perspective for understanding the cooperation between mathematicians that has become a pervasive aspect of mathematical research practice since the early twentieth century. The interesting corollary is that a different web of practices is a different conceptualization, which applies to mathematicians who have different areas of expertise. This will have implications when those areas may be interrelated in innovative ways, and mathematicians with different expertise cooperate.

References

1. Anacona, M., Arboleda, L.C y Pérez, F-J. 2014. On Bourbaki's axiomatic system for set theory. *Synthese* Vol. 191, 4069-4098.
2. Arrigoni, T. 2007. *What is meant by V? Reflections on the universe of all sets.* Mentis, Paderborn.
3. Benis Sinaceur, H. 2015. Is Dedekind a Logicist? Why does such a question arise? In M. Panza, G. Sandu, H. Benis Sinaceur, *Functions and Generality of Logic. Reflexions on Dedekind's and Frege's Logicisms.* Springer Verlag, 2015.
4. Bourbaki, N. 1950. 'The Architecture of Modern Mathematics'. In Ewald 1996, 1265–1276.
5. Bourbaki, N. 1954. *Théorie des ensembles. Éléments de Mathématique,* Première partie, Livre I, Chapitres I, II. Hermann & Cie, Paris.
6. Bourbaki, N. 1972. *Éléments d'histoire de Mathématique.* Hermann & Cie, Paris.
7. Chudnovsky, D., G. Chudnovsky, M. B. Nathanson 2004. *Number Theory: New York Seminar 2003.* Springer Verlag.
8. Corry, L. 1996. *Modern Algebra and the rise of mathematical structures.* Birkhäuser, Basel.
9. Corry, L. 1997. 'David Hilbert and the Axiomatization of Physics (1894–1905).' *Archive for History of the Exact Sciences* 51: 83–198
10. Dedekind, R. 1888. *Was sind und was sollen die Zahlen*? Braunschweig, Vieweg. References to the English version in Ewald 1996, 790–833.
11. Dirichlet, G. Lejeune 1837. Beweis des Satzes, dass jede unbegrenzte arithmetische Progression... unendlich viele Primzahlen enthält. *Abhandlungen der Königlich Preussischen Akademie der Wissenschaften* 1837, pp. 45–81. In *Dirichlet's Werke,* Vol. 1, ed. L. Kronecker. Berlin, Reimer, 1889, p. 313–342
12. Dirichlet, G. Lejeune and R. Dedekind 1871. *Vorlesungen über Zahlentheorie.* Braunschweig, Vieweg. English version ed. by AMS/LMS, 1991.
13. Ehrhardt, C. 2011. *Évariste Galois, la fabrication d'une icône mathématique.* Paris, Editions de l'Ecole Pratique de Hautes Etudes en Sciences Sociales.
14. Ewald, W. (ed.) 1996. *From Kant to Hilbert: A source book in the foundations of mathematics.* Vol. 2. Oxford Univ Press.
15. Ferreirós, J. 2005. R. Dedekind, *Was sind und was sollen die Zahlen?* (1888), G. Peano, *Arithmetices principia* (1889). In: *Landmark Writings in Western Mathematics 1640–1940,* ed. I. Grattan-Guinness (Elsevier, Amsterdam, 2005), 613–626.
16. Ferreirós, J. 2007. *Labyrinth of Thought: A history of set theory.* 2nd edn., Birkhäuser, Basel/Boston.
17. Ferreirós, J. 2007. The Early Development of Set Theory. *The Stanford Encyclopedia of Philosophy,* Edward N. Zalta (ed.), last revision 2020. https://plato.stanford.edu/archives/sum2020/entries/settheory-early/

18. Ferreirós, J. forthcoming. On Dedekind's Logicism. To appear in A. Arana & C. Alvarez (eds.), *Analytic Philosophy and the Foundations of Mathematics*. Palgrave Macmillan.
19. Ferreirós, J. 2016. *Mathematical Knowledge and the Interplay of Practices*. Princeton University Press.
20. Ferreirós, J. 2017. Dedekind's Map-theoretic Period. *Philosophia Mathematica* Vol 25:3, pp 318–340.
21. Ferreirós, J. and E. Reck 2020. Dedekind's Mathematical Structuralism. From Galois Theory to Numbers, Sets, and Functions. In E. Reck & G. Schiemer, eds. *The Prehistory of Mathematical Structuralism*. Oxford Univ. Press, 2020, 59–87.
22. Gauss, C. F. 1801. *Disquisitiones arithmeticae*. Leipzig, Fleischer. Repr. in *Werke*, Vol. 1, Leipzig, Teubner, 1863.
23. Gillies, D. A. 1982. *Frege, Dedekind, and Peano on the foundations of arithmetic*. Assen, Netherlands: Van Gorcum.
24. Goldstein, C. Schappacher, N. Schwermer, J. eds. 2006. *The Shaping of Arithmetic after C.F. Gauss's Disquisitiones Arithmeticae*. Springer Verlag.
25. Grassmann, H. G. 1861. *Lehrbuch der Mathematik*, Vol. I: *Arithmetik*. Berlin.
26. Hausdorff, F. 1914. *Grundzüge der Mengenlehre*, Leipzig, Veit. Repr. by Chelsea, NewYork, 1949.
27. Hilbert, D. 1917. Axiomatic Thinking. References to Ewald 1996, 1107–1115.
28. Hinkis, A. 2013. *Proofs of the Cantor-Bernstein Theorem—A Mathematical Excursion...* Birkhäuser, Basel.
29. James, I. M., ed. 1999. *History of Topology*. Amsterdam, Elsevier.
30. Kleiner, I. 2007. *A History of Abstract Algebra*. Basel/Boston, Birkhäuser.
31. Klev, A. M. 2017. Dedekind's Logicism. *Philosophia Mathematica*, Vol. 25:3, 341–368.
32. Kronecker, L. 1891. *'On the Concept of Number in Mathematics': Leopold Kronecker's 1891 Berlin Lectures,* ed. by J. Boniface & N. Schappacher. *Revue d'histoire des mathématiques* 7:2 (2001), page 207–277.
33. Kummer, E. E. 1975. *Collected Papers*, Vol. I: *Contributions to Number Theory*, ed. A.Weil. Berlin/New York: Springer-Verlag.
34. Mathias, A. R. D. 1992. The ignorance of Bourbaki. *The Mathematical Intelligencer* Vol 14, Issue 3, pp 4–13.
35. McLarty, C. 2006. Emmy Noether's set-theoretic topology. In Ferreirós & Gray, *The architecture of modern mathematics* (Oxford Univ Press, 2006), 187–208.
36. Neumann, P. M. ed. 2011. *The Mathematical Writings of Évariste Galois*. European Mathematical Society.
37. Peano, G. 1889. *Arithmetices principia, nova methodo exposita*. Torino. References to the partial English version in J. van Heijenoort, ed. *From Frege to Gödel* (Harvard UP, 1967), 85–97.
38. Peano, G. 1891. Sul concetto di numero. *Rivista di Matematica* 1: 87–102, 256–267. Repr. In *Opere Scelte*, Vol. 3 (Roma, Cremonese), 80–109.
39. Petri, B. Schappacher, N. 2006. On Arithmetization. In Goldstein, Schappacher & Schwermer 2006. Chapter V.2, 343–374.
40. Reck, E. H. 2013. Frege, Dedekind, and the Origins of Logicism. *History and Philosophy of Logic* Vol 34, issue 3: 242–265.
41. Scholz, E. 1999. The Concept of Manifold, 1850–1950. In James 1999, 25–64.
42. Segre, M. 1994. Peano's Axioms in their Historical Context, *Archive for History of Exact Sciences* 48: 201–342.
43. Skof, F. 2011. Giuseppe Peano between Mathematics and Logic. Springer Verlag.
44. Sieg, W. 2014 The Ways of Hilbert's Axiomatics: Structural and Formal. *Perspectives on Science* 22:1, 133–157.
45. Sieg, W. 2019. The Cantor–Bernstein theorem: how many proofs? *Philosophical Transactions of the Royal Soc. A* 377: 20180031.

46. Skolem, T. 1920. Logisch-kombinatorische Untersuchungen über die Erfüllbarkeit oder Beweisbarkeit mathematischer Sätze nebst einem Theoreme über dichte Mengen. *Videnskapsselskapet Skrifter, I. Matematisk-naturvidenskabelig Klasse*, **4**: 1–36. English version in J. van Heijenoort, ed., *From Frege to Gödel: A Source Book in Mathematical Logic, 1879–1931*, Harvard University Press, 252–263.
47. Stillwell, J. 1996. Introduction to Dedekind's *Theory of Algebraic Integers* (Cambridge Mathematical Library). Cambridge Univ. Press.
48. Van Heijenoort, J. 1967. From Frege to Gödel, 1879–1931. Harvard University Press.
49. Weber, H. 1893. Die allgemeinen Grundlagen der Galois'schen Gleichungstheorie. *Math. Annalen* **43,** 521–549.
50. Wussing, H. 1984. *The Genesis of the abstract group concept*. MIT Press.
51. Zermelo 1908. Neuer Beweis für die Möglichkeit einer Wohlordnung. Repr. in H.-D. Ebbinghaus, C. G. Fraser, A. Kanamori (eds.): *Ernst Zermelo. Collected Works / Gesammelte Werke*, Vol. 1 (Mengenlehre, Varia). Springer, Heidelberg, 2010.

José Ferreirós is professor of Logic and Philosophy of Science at the Universidad de Sevilla, Spain. A member of IMUS and of the *Académie Internationale de Philosophie des Sciences*, he was founding member of the APMP (Association for Philosophy of Mathematical Practices). Among his publications are *Labyrinth of Thought* (Birkhäuser, 1999), *Mathematical Knowledge and the Interplay of Practices* (Princeton UP, 2016), an intellectual biography of Riemann (*Riemanniana Selecta*, CSIC, Madrid, 2000), and the collective volume *The Architecture of Modern Mathematics* (Oxford UP, 2006). He has three daughters and loves music.

Chapter 7
Notes for a Seminar in Axiomatic Reasoning

Barry Mazur

Abstract This is not a standard article with *thesis, development, and conclusion*. It is rather a collection of notes and quotations meant to initiate reflections and discussions so that-guided by the passions and expertise of our students-my co-teachers (Amartya Sen and Eric Maskin) and I were able to shape our seminar course on *Axiomatic Reasoning*. I want to thank Giovanni Sommaruga for inviting me to include these notes in this volume, suggesting that I edit them slightly so that they offer a "many faceted view on axiomatics, something like a tour d'horizon of the subject."

7.1 Introduction to the Theme of Axiomatic Reasoning

I had written these notes in preparation for my part of a seminar course that I co-taught with Amartya Sen and Eric Maskin. The title of the course was *Axiomatic Reasoning* and it followed two other courses we three gave: *Reasoning via Models* and *Utility*; and it preceded a course entitled *Subjectivity/Objectivity*.

I want to thank Giovanni Sommaruga for inviting me to include these notes in this volume, suggesting that I edit them slightly so that they offer a "many faceted view on axiomatics, something like a tour d'horizon of the subject."

This, then, is not a standard article with *thesis, development, and conclusion*. It is rather a collection of (slightly edited) notes and quotations meant as 'starter' to precede course readings, and to initiate reflections and discussions; and perhaps eventually to connect with the projects of students. It was quite a joy to teach with Amartya and Eric—and to engage with our students. The intent of our four courses

Thanks so much to the students of our class, and to Mikhail Katz, for a close reading of these notes and for extremely helpful comments.

B. Mazur (✉)
Department of Mathematics, Harvard University, One Oxford Street, Cambridge, MA 02138-2901, USA
e-mail: mazur@math.harvard.edu

F. Ferreira et al. (eds.), *Axiomatic Thinking I*,
https://doi.org/10.1007/978-3-030-77657-2_7

was to create an opportunity for ourselves and our students to *live with* a concept for a significant length of time—without specifying a particular goal other than to become at home with, intimate with, the concept in broad terms in its various facts and its various moods. To become acquainted with a bit of the history of the concept, its reception, its development.

Such an experience can provide resonances which enrich thoughts that one may have, or can connect with ideas that one encounters, years later.[1] Our aim was to shape our seminar following the different experiences, background, viewpoints, and preferences of our students.

So these notes constituted my share of invitations to our students to help them choose directions of interest to them—which in concert with those of my co-teachers—would fashion the thrust of our course.

7.1.1 Axioms, Taken Broadly

The etymological root of the word *axiom* is the Greek αξιωμα meaning 'what is fitting.' The concept *axiom* is often taken to simply mean 'self-evident assertion,' but we will take a much broader view, allowing it to encompass frame-creating assertions ranging from 'common notions' (Euclid's favorite) to 'rules,' 'postulates,' 'hypotheses,' and even 'definitions' and 'characterizations' if they play a suitable role in the ensuing discussion.

The proclamation of an 'axiom' as *self-evident* calls, at bottom, for a coherent agreement of belief of that assertion by all humanity.[2]

One can only be reminded, though, of the fragility of such coherent agreement, by considering the claim in the United States Declaration of Independence:

> We hold these Truths to be *self-evident*, that all Men are created equal, that they are endowed by their Creator with certain unalienable Rights, that among these are Life, Liberty, and the Pursuit of Happiness...

[1] This follows the format of some seminars I once taught with the late historian of Science John Murdoch. Murdoch would ask me at the beginning of the year: "What do you want to know?" and he would shape the readings and theme of the ensuing course often based on my answer. This "What do *you* want to *know*", being abruptly personal, allowed the seminars to be somewhat oblique to the standard professional academic themes of discourse.

[2] But there is St. Thomas Acquinas's view in *Summa Theologiae*: "... something can be self-evident in two ways; in one way, in itself but not relative to us; in the other way, both in itself and relative to us. For what makes a proposition self-evident is the fact that its predicate is included in the concept of its subject, e.g., *man is an animal*, for animal is included in the concept man. ... I say that the proposition "God exists" is self-evident when considered in itself, because its predicate and subject are identical, for God is his existence. ... But because we do not know about God what he is, the aforementioned proposition is not self-evident to us, but it needs to be demonstrated by means of those things that are more known relative to us although less known in themselves, namely, by means of his effects." St. Thomas Aquinas, Summa Theologiae, Part One, q. 2 (transl.: Gerald J. Massey); for the text go to: http://www.pitt.edu/~gmas/1080/Q2A1.htm.

and its revision—demoting "self-evident" to "proposition"[3]—by Abraham Lincoln, under the pressure of the American Civil War:

> Four score and seven years ago our fathers brought forth, on this continent, a new nation, conceived in Liberty, and dedicated to the *proposition* that all men are created equal.

Axioms are a fundamental tool, a way of grounding a reasoned argument, a way of making explicit one's 'priors' or prior assumptions, a way of stipulating assertions[4] very clearly so as to investigate their consequences, of organizing beliefs; a way of... in short, reasoning.

Axiomatic frameworks offer striking transparency and help open to view the lurking assumptions and presumptions that might otherwise be unacknowledged. This mode of thought has been with us at least since Aristotle.

Axioms in formal (and even sometimes in somewhat informal) structures constitute an 'MO' of mathematics at least since Euclid, but surely earlier as well. "Surely," despite, curiously, the lack of any earlier record of it; and despite the fact that there is substantial record of much earlier mathematical thought. Egyptian mathematical papyri contain quite an array of problems and their solutions—e.g., the Moscow Mathematical Papyrus, dated approximately 1700 BC offers a correct discussion regarding the volume of a truncated square pyramid and a step-by-step computation of a particular example—but nowhere in these papyri is there as much as a hint of any mathematical protocol for demonstration, let alone any axiomatic foundational structure.

We will see how the very core of meaning and use of *axiom* in mathematics has undergone quite an evolution, through Euclid, his later commentators, Hilbert's revision of the notion of axiom, and the more contemporary set theorists.

Axioms are standard structures as they appear in models in the sciences, sometimes occurring as proclaimed 'laws': borrowing that word from its *legal* roots. Newton's Laws act as axioms for Classical Mechanics, the fundamental laws of thermodynamics for Thermodynamics.

Similarly for Economics: Axiomatic Utility Theory is very well named where the 'axioms' play more the role of *desiderata* which may or may not be realizable, especially in the face of the variously named 'paradoxes' and 'Impossibility Theorems' (as Professors Maskin and Sen will be discussing). Nevertheless this axiomatic format provides us with an enormously useful and powerful tool to understand forces at play in Economics.

Noam Chomsky's *Structural linguistics* set the stage for a very axiomatic approach to language acquisition and use (with interesting later critique) as does the vast tradition of rule-based grammars (as Professor Sen will be discussing later in this seminar).

---

[3] i.e., something that has to be proved.

[4] Sometimes formulated in the spirit of the notion of "stipulation" analogous to its legal use (which is a formal agreement made between opposing parties before a pending hearing or trial, that some specific statement—which might otherwise be taken as debatable—is to be taken as correct and beyond debate).

Rules in games, and in the formal set-ups in mathematical game theory have their distinct qualities. Even more of 'distinct quality' is the subtle manner in which it is sometimes understood that rules are *not* expected to be strictly obeyed; e.g., as in the composition of a sonnet.

In the Bayesian mode of inductive reasoning, the 'priors' (as the Bayesians call them)—which are, in effect, input axioms—are constantly re-assessed in connection with the flow of further incoming data ("the data educates the priors" as they sometimes say). This is also quite a distinctive way of dealing with one's axioms!

Model Theory focusing attention, as it does, on language and syntax rather than semantics (in fact, making syntax the initiating substance of its investigations) offers an interesting perspective on the notion of axiom. And Set Theory, with the extraordinary history of the axiomatic crises related to this subject—and with the role that sets often play as the substrate of axomatic systems—deserves our attention too.

We live these days at a time when computer programs, governed by *algorithms*—hence a specific form of axiomatic reasoning—make selections for us, recommendations, choices, and sometimes critical decisions. The question of when, and how, more flexible modes of human judgment should combine with, and possibly mitigate, axiom-driven decision processes is a daily concern, and something that we might address, at least a bit, in our seminar.[5]

Most curiously, axiomatic structure has come up in various reflections regarding moral issues. (This, by the way, happens more than merely in the *golden rule*: we might read a bit of Spinoza's *Ethics* which is set up in the formal mode of mathematical discourse, complete with Postulates, Definitions, and Theorems.)

7.2 The Evolution of Definitions and Axioms, from Ancient Greek Philosophy and Mathematics to Hilbert

7.2.1 Hypotheses, Definitions, Premises Before Euclid

Here is Socrates lecturing to Adeimantus in Plato's *Republic* VI.510c, d:

> *...the men who work in geometry, calculation, and the like treat as known the odd and the even, the figures, three forms of angles, and other things akin to these in each kind of inquiry.*
>
> *These things they make* **hypotheses** *and don't think it worthwhile to give any further account of them to themselves or others as though they were clear to all. Beginning from them, they ... make the arguments for the sake of the square itself and the diagonal itself, not for the sake of the diagonal they draw, and likewise with the rest. These things themselves that they mold and draw—shadows and images in water—they now use as images, seeking to see those things themselves, that one can see in no other way than with thought.*

[5] Relevant to this discussion is Stephen Wolfram's *A New Kind of Science* (https://www.wolframscience.com/) in which it is proposed that one simply replace equations in scientific laws with algorithms.

So the term 'hypothesis,'[6] here as elsewhere in the platonic dialogues, means *starting points in arguments*—starting points assessed to have no need for prior explanation, or justification. After all, a rational argument has to start somewhere. Hypotheses taken in this sense—starters—might be regarded as proto-axioms.

Nevertheless, in Book VII of the *Republic*,[7] Plato presses deeper:

> Is not dialectics the only process of inquiry that advances in this manner, doing away with hypotheses, up to the first principle itself in order to find confirmation there?

"Confirmation," then, is what is often referred to as his theory of 'forms', or the εἰδε. So Plato—unsatisfied, it seems, with mere hypotheses— is suggesting a 'process of inquiry' that seeks to go deeper (or higher) than unsubstantiated hypotheses (of mathematics, or more generally, of thought) to find their grounding in another level: "the first principle itself."

In Aristotle we find a jump to quite a different meta-level: *organization*. Organizational schemes of logic, such as the *Organon* of Aristotle,[8] have been vastly influential and have been—even if largely implicit—the armature of the way in which we formulate assertions, ask questions, and reach conclusions in mathematics as in everything else. Aristotle begins his discussion in the *Prior Analytics* by setting for himself quite a task: to pin down *demonstration* "and for the sake of demonstrative science," to:

> ... define a premise, a term, and a syllogism, [*and the nature of a perfect and of an imperfect syllogism; and after that, the inclusion or non-inclusion of one term in another as in a whole, and what we mean by predicating one term of all, or none, of another...*]

Aristotle gets to this job right away, but Aristotelian commentarists (including Alexander of Aphrodiasis[9] and Philiponus[10]) have much to say about the difficulty of interpretation of the basic ingredients of the sentence just quoted. For example, syllogism (συλλογισμός) might be taken to mean, broadly, *deduction*. For Aristotle characterizes *syllogism* as:

> discourse in which, certain things being stated, something other than what is stated follows of necessity from their being so.

We might take Aristotle, here, to be contemplating a meta-logical overview of rational discourse rather than have the word syllogism signify (with formal logical specificity) what that word means nowadays.

---

[6] The etymological breakdown of the Greek word would suggest the action of "putting or placing under".

[7] See http://www.perseus.tufts.edu/hopper/ Republic VII 533a–d.

[8] Aristotle: Prior Analytics (Chap. 1 of Book I) https://en.wikisource.org/wiki/Prior_Analytics.

[9] See p. 189 of *Natural and Artifactual Objects in Contemporary Metaphysics* edited by Richard Davies, Bloomsbury.

[10] For an overview on discussions regarding the meaning of *perfect* and imperfect syllogisms, see *A Rationale for Aristotle's Notion of Perfect Syllogisms* Kevin L. Flannery, Notre Dame Journal of Formal Logic **28** (1987) https://projecteuclid.org/download/pdf_1/euclid.ndjfl/1093637566.

Aristotle offers this succinct definition of *premise* (πρότασις) neatly distinguishing between those propositions (λόγοι) involving universal, no, or existential quantification:

> A premise then is a 'proposition' which affirms or denies something of something, and this is universal, or particular, or indefinite.

I put scare-quotes around 'proposition' since I'm imagining πρότασις as also being used here to encompass the subject of this seminar; i.e., axiom.[11]

The contemporary view of formal logic (e.g., first-order logic) owes much to Aristotle's formulation.

Since *definition*, defined by Aristotle as an:

> *account which signifies what it is* **to be** *for something.*[12]

plays such a vital role in mathematics, the notion deserves close attention. Mathematics seems to require as strict lack-of-ambiguity in its assertions as possible, and therefore maximal clarity in its definitions. But perhaps—since ambiguity is sometimes unavoidable—it is better to say that any ambiguity should be unambiguously labeled as such.

The nature, and role, of *definition* in mathematical usage has evolved in remarkable ways. We will be discussing this in more detail later, but consider the first two definitions in Book I of Euclid's *Elements*[13]:

(i) A point is that which has no part.
(ii) A line is breadthless length.

and their counterparts in Hilbert's rewriting of Euclid's *Elements*.

Hilbert's *Foundations of Geometry*[14] begins with:

> Let us consider three distinct systems of things. The things composing the first system, we will call **points** and designate them by the letters A, B, C,. . . ; those of the second, we will call **straight lines** and designate them by the letters a, b, c,. . . The points are called the **elements of linear geometry**; the points and straight lines, **the elements of plane geometry**. . .

---

[11] For a full discussion of this Aristotelian vocabulary, see *Aristotle's Logic* (Stanford Encyclopedia of Philosophy) https://plato.stanford.edu/entries/aristotle-logic/.

[12] A puzzling definition: *logos ho to ti ên einai sêmainei*.

[13] These 'Elements' have quite an impressive spread, starting with the proclamation that a point is characterized by the property of 'having no part,' and ending with its last three books, deep into the geometry of solids, their volumes, and the five Platonic solids. It is tempting to interpret this choice of ending for the *Elements* as something of a response to the curious interchange between Socrates and Glaucon in Plato's *Republic* (528a–d) where the issue was whether Solid Geometry should precede Astronomy, and whether the mathematicians had messed things up.

It also would be great to know exactly how—in contrast—the *Elements* of Hippocrates of Chios ended. (It was written over a century before Euclid's *Elements* but, unfortunately, has been lost.)

[14] David Hilbert, *Foundations of Geometry* (trans. E. J. Townsend), The Open Court Publishing Co. (1950); for the text go to: https://www.gutenberg.org/files/17384/17384-pdf.pdf.

Hilbert's undefined terms are: *point, line, plane, lie, between,* and *congruence.*

One might call Euclid's and Hilbert's formulations **primordial definitions** since they spring ab ovo–i.e., from nothing. Or at least from 'things' not in the formalized arena of mathematics, such as Hilbert's "*system of things*". Euclid's definitions of *point* and *line* seem to be whittling these concepts into their pure form from some more materially graspable context[15] (e.g., where lines have breadth) while for Hilbert the essence of *point* and *line* is their relationship one to the other.

Once one allows *Sets* to be the underlying material of some of the basic mathematical concepts—i.e., once one accepts the bedrock of Set Theory, definitions often have the form of being 'delineations of structure,' cut out by means of quantifiers and predicates but making use of set theoretic, or other priorly defined objects. E.g. *A circle is a set of points equi-distant from a single point in the Euclidean plane.*

We will discuss all this in a moment.

The essential roles that 'definition' play for us are: to delineate the objects of interest to be studied; to encapsulate; to abbreviate; and to focus.

7.2.2 *The Axiomatic 'Method'*

Axioms, as we've seen, have been around—at least—since ancient Greek mathematical activity, but only more recently have people viewed the act of 'listing axioms' as a *method,* rather than (somewhat more relaxedly) as a natural move to help systematize thought.

It may have been David Hilbert who actually introduced the phrase *axiomatic thinking* to signal the fundamental role that the structure of an axiomatic system plays in mathematics. Hilbert clearly views himself as molding a somewhat new architecture of mathematical organization in his 1918 article "Axiomatisches Denken" (Mathematische Annalen 78, 405–415). The English translation begins[16] with a political metaphor, that neighboring sciences being like neighboring nations need excellent internal order, but also good relations one with another, and:

> ... The essence of these relations and the ground of their fertility will be explained, I believe, if I sketch to you that general method of inquiry which appears to grow more and more significant in modern mathematics; the axiomatic method, I mean.

Hilbert's essay ends:

> In conclusion, I should like to summarize my general understanding of the axiomatic method in a few lines. I believe: **Everything that can be the object of scientific thinking in general, as soon as it is ripe to be formulated as a theory, runs into the axiomatic method and thereby indirectly to mathematics.** Forging ahead towards the ever deeper layers of axioms in the above sense we attain ever deepening insights into the essence of scientific thinking itself, and we become ever more clearly conscious of the unity of our knowledge. In the

[15] I want to thank Eva Brann for pointing this out.

[16] AXIOMATIC THINKING, Philosophia Mathematica, Volumes 1–7, Issue 1–2, June 1970, pp. 1–12; go to: https://doi.org/10.1093/philmat/s1-7.1-2.1.

evidence of the axiomatic method, it seems, mathematics is summoned to play a leading role in science in general.

7.2.3 Definitions and 'Characterizations'

In the "Definitions" of Euclid's *Elements* it is striking how these are both definitions (or at the very least descriptions) as well as axioms. Definitions, of course, suffer the risk of being 'ambiguous' and collections of axioms suffer the risk of being inconsistent or—in various ways–inadequate. The essential roles that 'definition' play for us are: to delineate the objects of interest to be studied; to encapsulate; to abbreviate; and to focus.

As for the power of definition to provide 'focus,' consider the distinction between *definition* and *characterization*—as in the two equivalent definitions of prime number (given by (i) and (ii) below)—where one is left to make the choice of regarding one of these as 'definition' and the other as 'characterization':

A prime number p is a (whole) number greater than one

(i) that is not expressible as the product of two smaller numbers.

or

(ii) having the property that if it divides a product of two smaller numbers, it divides one of them.

If you choose (ii) as the fundamental definition you are placing the notion of prime number in the broader context of 'prime'-ness as it applies to number systems more general than the ring of ordinary numbers—and more specifically in the context of *prime ideals* of a general ring. So choosing (ii) as definition casts (i) as a specific feature that characterizes prime numbers, thanks to the theorem that guarantees the equivalence of these to formulations. Going the other route—i.e., focusing on (i), the unfactorable quality of prime number, would then cast (ii) as a basic more general feature also *characterizing* prime-ness.

7.2.4 The Adequacy or Inadequacy of Axioms

Even when axioms seem inadequate, or fail, the discussion can continue in an interesting way—very often dealing with more subjective issues. As Gödel's Incompleteness Theorem (see the discussion below) points to the striking limitations of Hilbert's grand notion of *formal system*, these limitations themselves have interesting implications. Similarly, vastly illuminating are the limitations implied by Arrow's Impossibility Theorem regarding social choice theory; or the "named paradoxes" connected to axiomatic utility theory—as presented by my co-teachers in our seminar—paradoxes that seem to suggest that choices in certain situations violate various formulated axioms that were meant to be in track with expected rational behavior; paradoxes such as:

- *Allais's Paradox*
  The 'paradox' is that there are situations where the (ostensible) adding of 'equals' (i.e., adding further equal alternatives) to two choices that are open to us gets us to switch our preferred choice. The issue motivating that switch is that a *guaranteed* very large winning beats—in our assessment—a possible quite greater winning but with a 1% probability of total loss.[17]

or

- *Ellsberg's Paradox*
  The 'paradox' is related to a kind of *meta-risk assessment;* knowing clearly what the odds are beats having the odds a bit indeterminate. Ellsberg sets up a game **I** where you must make a choice (A or B). And then he compares this game to a modified version **II** with slightly different choices (A' or B'). In the first game **I** you actually know—i.e., can reasonably compute—the odds of winning depending on your choice; and you would most likely choose A rather than B. Ellsberg then changes the game by modifying the two choices A or B *but: in an equal way!* so that in game **II** you are faced with choices A' or B' but now you don't quite know the odds; and... curiously... you would now likely choose B'.[18]

or

- *the St. Petersburg Paradox* and the (much later) ideas of Kahnemann-Tversky—that emphasize the deeply subjective nature—and intertwining—of *utility* and *expectation.*[19]

7.2.5 The Evolution of the Form and Role of Axioms

The variants and evolution of the notion of axiom, the changes in formulation and use, is striking. Compare the different axiomatic formats of Euclidean geometry, as conceived in what we might categorize as formulated

- *synthetically* by Euclid, David Hilbert,
- sort of *synthetically* by George Birkhoff,
- *analytically* by Descartes,[20]
- *holistically*—one might say—which is the viewpoint of the 'Erlangen Program' at least as it connects with the formulation of Euclidean Geometry (we'll only briefly discuss this).

---

[17] (See the Wikipedia entry: https://en.wikipedia.org/wiki/Allais_paradox).

[18] (See the Wikipedia entry: https://en.wikipedia.org/wiki/Ellsberg_paradox).

[19] (See: https://plato.stanford.edu/entries/paradox-stpetersburg/).

[20] And there is a hint of a beginning analytic view in Nicholas of Oresme's *Geometry of Qualities*. See Sect. 7.2.6 below.

7.2.5.1 Euclid's Elements, Book I

Euclid begins his tour through the *Elements*[21] labeling his starting assertions as *Definitions, Postulates, or Common Notions.*

7.2.5.2 Euclid's Definitions

His *Definitions* are meant to acquaint you with—or in certain cases, just to name—some basic creatures of his 'Euclidean Plane:' point, line, straight line, extremity of line, surface, boundary, circle, etc. How curious these formulations are, given that they are, after all, formal definitions meant to be helpful, and explicitly used, in logical arguments. As we run through them we might discuss exactly how useful each of these definitions could possibly be in a formal argument:

(i) A **point** is that which has no part.
(ii) A **line** is breadthless length.
(iii) The extremities of a line are points.
(iv) A **straight line** is a line which lies evenly with the points on itself.
The 'familiar' corresponding definitions, not at all in the spirit of these definitions appear in later interpretations of Euclid[22]:

> *A straight line is* [implied: **uniquely**] *determined by two of its points.*
>
> *Or:*
>
> *A straight line segment is the* [implied: **unique**] *curve of shortest distance between its endpoints.*

(v) A **surface** is that which has length and breadth only.
(vi) The **extremities of a surface** are lines.
(vii) A **plane surface** is a surface which lies evenly with the straight lines on itself.
(viii) A **plane angle** is the inclination to one another of two lines in a plane which meet one another and do not lie in a straight line.
(ix) And when the lines containing the angle are straight, the angle is called **rectilineal**.
(x) When a straight line set up on a straight line makes the adjacent angles equal to one another, each of the equal angles is **right,** and the straight line standing on the other is called a **perpendicular** to that on which it stands.

---

[21] Sir Thomas L. Heath/Euclid: The Thirteen Books of The Elements/Volume 1/Second Edition.
For an (online) interlinear Greek-English edition of Euclid's Elements—*the Greek text of J.L. Heiberg (1883–1885) and an English translation, by Richard Fitzpatrick*—go to: http://farside.ph.utexas.edu/Books/Euclid/Elements.pdf.

[22] For example, see T. L. Heath's discussion of Proclus's and Archimedes' take on this definition, in the footnote on pp. 3 and 4 of his edition (and translation) of Archimedes' *On the Sphere and Cylinder I* in *The Works of Archimedes* Dover (2002); for the text go to https://math.mit.edu/~mqt/math/resources/archimedes_on-the-sphere-and-the-cylinder_bk-1.pdf.

The next two definitions introduce–in passage, so to speak—an order relation between angles (but nothing explicit is mentioned about the nature of this order relation):

(xi) An **obtuse angle** is an angle greater than a right angle.
(xii) An **acute angle** is an angle less than a right angle.
(xiii) A **boundary** is that which is an extremity of anything.
(xiv) A **figure** is that which is contained by any boundary or boundaries.
(xv) A **circle** is a plane figure contained by one line such that all the straight lines falling upon it from one point among those lying within the figure are equal to one another;

The next definition includes—in passage—the claim that **the** *"one point" described in Definition* (xv) *above is uniquely characterized by its properties given in that definition:*

(xvi) And the point is called the **center of the circle**.

In the next definition the final sentence would seem to follow from (xv):

(xvii) A **diameter of the circle** is any straight line drawn through the center and terminated in both directions by the circumference of the circle. And such a straight line also bisects the center.
(xviii) A **semicircle** is the figure contained by the diameter and the circumference cut off by it. And the center of the semicircle is the same as that of the circle.

The remaining definitions offer straightforward terminology for structures related to concepts already defined:

(xix) **Rectilineal figures** are those which are contained by straight lines, **trilateral figures** being those contained by three, quadrilateral those contained by four, and multi-lateral those contained by more than four straight lines.
(xx) Of trilateral figures, an **equilateral triangle** is that which has its three sides equal, an **isosceles triangle** that which has two of its sides alone equal, and a **scalene triangle** that which has its three sides unequal.
(xxi) Further, of trilateral figures, a **right-angled triangle** is that which has a right angle, an **obtuse-angled triangle** that which has an obtuse angle, and an **acute-angled triangle** that which has its three angles acute.
(xxii) Of quadrilateral figures, a **square** is that which is both equilateral and right-angled; an **oblong** that which is right-angled but not equilateral; a **rhombus** that which is equilateral but not right-angled; and a **rhomboid** that which has its opposite sides equal to one another but is neither equilateral nor right-angled. And let quadrilaterals other than these be called **trapezia**.
(xxiii) **Parallel straight lines** are straight lines which, being in the same plane and being produced indefinitely in either direction, do not meet one another in either direction.

7.2.5.3 Euclid's Postulates

It is a bit more difficult to give a single word description of the nature of his *Postulates*. The first three are descriptions of (the axiomatically allowed) constructions. The emphasis is on construction rather than existence. The fourth has the function of defining 'equality' of (right) angles, which opens a topic for discussion that I'll return to below. The fifth is indeed the celebrated "Fifth Postulate", whose status was of interest to the earliest commentators, and whose ramifications weren't adequately understood until the beginning of the 19th century. The major issue is the question of the dependence of this fifth postulate on the rest of the axiomatic set-up; the minute one questions its independence, one is on the way to model-formation.

(i) To draw a straight line from any point to any point.
(ii) To produce a finite straight line continuously in a straight line.
(iii) To describe a circle with any centre and distance.
(iv) That all right angles are equal to one another.
(v) ('Fifth Postulate':) That, if a straight line falling on two straight lines make the interior angles on the same side less than two right angles, the two straight lines, if produced indefinitely, meet on that side on which are the angles less than the two right angles.

7.2.5.4 Euclid's Common Notions

These are closest to modern axiomatics, formulating rules regarding the terms *equal* and *greater than*. But the intuitive notion—that two angles are *equal* if there is a ("Euclidean") transformation bringing one exactly onto the other—is utterly absent from Euclid's vocabulary—despite the fact that it is highly suggested (by (iv) below).

(i) Things which are equal to the same thing are also equal to one another.
(ii) If equals be added to equals, the wholes are equal.
(iii) If equals be subtracted from equals, the remainders are equal.
(iv) Things which coincide with one another are equal to one another.
(v) The whole is greater than the part.

Remark 1 Viewing this axiom structure, three issues stand out:

- **Issues of uniqueness** are perhaps implied (both in Euclid, and Archimedes, below) but not specifically mentioned.
- **Substrate**: The most striking fact about these definitions is that they don't rely on set theoretic vocabulary. We moderns immediately think 'sets,' 'subsets,' 'membership in sets,' etc. and tend to build our structures starting with sets as substrate, as in Hilbert's axioms below.
- **Motion**: Euclid has no vocabulary *at all* for 'continuous motion,' 'transformation,' 'function' except as these issues are introduced in the *Postulates* and/or when one triangle is "applied" to another. This is in contrast to, say, *'Erlangen Program'* which we will discuss below.

7.2.5.5 Archimedes, *On the Sphere and Cylinder I*

In the treatise *On the Sphere and Cylinder*[23] Archimedes begins in a somewhat personal way:

> Archimedes to Dositheus, greeting:

Archimedes goes on to say that on a former occasion he had sent Dositheus some results, and now wants to send him some more results. Archimedes lists five 'assumptions'—i.e., in effect: axioms—that he will depend upon. This, then, isn't an organized discursive structure such as Euclid's but implicitly depends upon Euclid, and on that list of supplementary assumptions. These assumptions are all provable with the tools of multivariable calculus, based on the standard (Cartesian) analytic foundations of Euclidean geometry. The posing of these "assumptions" are not for establishing a fundamental axiomatic framework to be the grounding of the progress of Geometry as a whole; but this collection of assumptions are put together to perform the specific aim that is the focus of this particular treatise.

(i) Of all the lines which have the same extremities the straight line is the least.
(ii) Of other lines in a plane and having the same extremities any two of them are unequal if they are both concave in the same direction and one is between the other and the straight line with the same extremities.
(iii) ... similar to (i) but defining plane among surfaces with the same 'extremities'.
(iv) ... similar to (ii) but distinguishing surfaces among surfaces with the same 'extremities'.
(v) "*Archimedes Principle*": Further of unequal lines, unequal planes, unequal solids, the greater exceeds the lesser by such a magnitude that when added to itself, can be made to exceed any assigned magnitude among those which are comparable...

7.2.6 *The Unfolding of Analytic Geometry*

7.2.6.1 Nicholas of Oresme, The Geometry of Qualities

Nicholas of Oresme's 14th century treatise[24] already envisioning the geometry (and the utility) of what we call graphs, begins:

> On the Continuity of Intensity

---

[23] T. L. Heath (ed. and transl.) Archimedes' *On the Sphere and Cylinder I* in *The Works of Archimedes* Dover (2002); for the text go to https://math.mit.edu/~mqt/math/resources/archimedes_on-the-sphere-and-the-cylinder_bk-1.pdf.

[24] *De configurationibus qualitatum et motuum*. See *Nicole Oresme and the Geometry of Qualities and Motions* Marshall Clagett, University of Wisconsin Press (1968).

> Every measurable thing except numbers is imagined in the manner of continuous quantity. Therefore, for the mensuration of such a thing, it is necessary that points, lines, and surfaces, or their properties, be imagined.

What ensues is a discussion of a manner of graphically illustrating 'qualities.' What we would call "the x-axis," Oresme calls "the subject". Oresme reserves what we would call the y-axis for what he calls the "intensity, or the longitude, of quality". For a given subject and quality, Oresme would erect a perpendicular line emanating from each point x of the subject—the line being in proportion to the intensity of the quality at the point x of the subject. For example, if the "subject" is time and the "quality" is distance from home, the "configuration," i.e., the graph—something we moderns are utterly familiar with—would offer for Oresme a 'configuration of motion,' and he devotes a section of his treatise to "the beauty of configurations of velocities," *de pulchritudine configurationum velocitatum*. Oresme allows for a much wider assortment of possible subjects—quite wild—and I won't spoil it for people who haven't yet read his treatise by listing them.

7.2.6.2 René Descartes, *La Géométrie*

Oresme's treatise was written about two centuries earlier than the much more familiar 'Cartesian Geometry'—originating in René Descartes' *La Géométrie,*[25] the last of three appendices of his *Discours de la méthode* (1637). It begins:

> Des problèmes qu'on peut construire sans y employer que des cercles et des lignes droites.
>
> Tous les problèmes de géométrie se peuvent facilement réduire à tels termes, qu'il n'est besoin par après que de connoître la longueur de quelques lignes droites pour les construire.[26]

(Problems the construction of which uses only circles and straight lines.

All geometrical problems may be easily reduced to such terms that afterwards one only needs to know the lengths of certain straight lines in order to construct them.)

This, then, is the starter for **modern analytic geometry**, where:

- A **point** is (given by) a couple (x, y) where x and y are real numbers; i.e., the **plane** is (given by) the 2-dimensional vector space $\mathbf{R}^2$ over the field $\mathbf{R}$ of real numbers.
- A **line** is the locus of a "linear equation"

$$y = ax + b$$

where $a, b \in \mathbf{R}$.
- etc.. . .

In a sense Birkhoff's axiom system (see Sect. 7.2.7.3 below) is something of a hybrid synthetic/analytic axiomatic set-up.

---

[25] Oeuvres de Descartes, éd. Cousin, tome V.

[26] It continues: "Et comme toute l'arithmétique n'est composée que de quatre ou cinq opérations, qui sont, l'addition, la soustraction, la multiplication, la division, et l'extraction des racines, qu'on peut prendre pour une espèce de division, ainsi n'a-t-on autre chose à faire en géométrie touchant les lignes qu'on cherche pour les préparer à être connues, que leur en ajouter d'autres, ou en ôter. . . ".

7.2.7 *Modern Ways of Interpreting Euclid's Axioms*

7.2.7.1 Hilbert's Euclidean Geometry

Hilbert's axiom system[27] is constructed with six **primitive notions:**

(i) three primitive terms: *point, line, plane*, and
(ii) three primitive relations:

- *Betweenness,* a ternary relation linking points;
- *Lies on (Containment)*, three binary relations, one linking points and straight lines, one linking points and planes, and one linking straight lines and planes;
- *Congruence*, two binary relations, one linking line segments and one linking angles. Note that line segments, angles, and triangles may each be defined in terms of points and straight lines, using the relations of betweenness and containment. All points, straight lines, and planes in the following axioms are distinct unless otherwise stated.

And there are these structures, axioms, and lists of "defined terms":

(iii) **Incidence**: For every two points A and B there exists a line a that contains them both...
(iv) **Order**: If a point B lies between points A and C, B is also between C and A, and there exists a line containing the distinct points A,B,C ... If A and C are two points of a line, then there exists at least one point B lying between A and C. Of any three points situated on a line, there is no more than one which lies between the other two.
(v) **Pasch's Axiom**: Let A, B, C be three points not lying in the same line and let L be a line lying in the plane ABC and not passing through any of the points A, B, C. Then, if the line L passes through a point of the segment AB, it will also pass through either a point of the segment BC or a point of the segment AC.
(vi) **Axiom of Parallels**: Let m be any line and A be a point not on it. Then there is a unique line in the plane, determined by m and A, that passes through A and does not intersect m.
(vii) **Congruence**: If A, B are two points on a line L, and if A' is a point upon the same or another line L', then, upon a given side of A' on the straight line L', we can always find a point B' so that the segment AB is congruent to the segment $A'B$ ...
(viii) **Continuity**:

- *Axiom of Archimedes*: If AB and CD are any segments then there exists a number n such that n segments CD constructed contiguously from A, along the ray from A through B, will pass beyond the point B.

[27] See footnote 7.2.1 above; or: http://web.mnstate.edu/peil/geometry/C2EuclidNonEuclid/hilberts.htm.

- *Axiom of line completeness*: An extension of a set of points on a line with its order and congruence relations that would preserve the relations existing among the original elements as well as the fundamental properties of line order and congruence that follows from [the axioms discussed] is impossible.

(ix) **Defined Terms**: *segment, ray, interior, triangle, 'lie on the same side', ...*

7.2.7.2 Remarks About Hilbert's Axioms

- Hilbert's Axioms offer an articulation very different from Euclid's: the triple *definitions/postulates/common notions* being replaced by *primitive terms/primitive relations/structures and axioms.*
- The common notions (i.e., logical pre-structures like 'equality') are implicitly assumed rather than formulated.
- Modern quantification is explicit. E.g., the 'Incidence Axiom' calls up universal and existential quantification: $\forall$ points A, B, $\exists$ a line through A and B.
- **Set Theory**: Most importantly, Hilbert expresses his axioms in Set theoretic vocabulary.
  But if one uses Set Theory as a 'substrate' on which to build the structures of mathematics, as in the classical *Grundlagen der Mathematik* of Bernays and Hilbert, one must tangle with all the definitional questions that are faced by Set Theory (starting with: *what is a set?* and continuing with the discussion generated by the work of Frege, Russell, etc.).[28]
- **Infinity**. And then compare all this with the discussion about the existence of infinite sets in Bernays-Hilbert's *Grundlagen der Mathematik, Vol. I*:

> ... reference to non-mathematical objects can not settle the question whether an infinite manifold exists; the question must be solved within mathematics itself. But how should one make a start with such a solution? At first glance it seems that something impossible is being demanded here: to present infinitely many individuals is impossible in principle; therefore **an infinite domain of individuals as such can only be indicated through its structure, i.e., through relations holding among its elements.** In other words: a proof must be given that for this domain certain formal relations can be satisfied. The existence of an infinite domain of individuals can not be represented in any other way than through the satisfiability of certain logical formulas...

From Hilbert's *On The Infinite*[29]:

> ... We encounter a completely different and quite unique conception of the notion of infinity in the important and fruitful method of ideal elements. The method of ideal

[28] For example, go back to Dedekind's marvelous idea of capturing the notion of *infinite* by discussing self-maps (this notion popularized by people checking into Hilbert's hotel). You might formulate Dedekind's idea this way: a set S is **infinite** if it admits an injective but non-surjective self-map... and then confuse yourself by trying to figure out how this compares with the property that S admits a surjective but non-injective self-map.

[29] See: Paul Benacerraf, Hilary Putnam *Philosophy of Mathematics*, Cambridge University Press (1984) 183–201.

elements is used even in elementary plane geometry. *The points and straight lines of the plane originally are real, actually existent objects.* One of the axioms that hold for them is the axiom of connection: one and only one straight line passes through two points. It follows from this axiom that two straight lines intersect at most at one point. There is no theorem that two straight lines always intersect at some point, however, for the two straight lines might well be parallel. Still we know that by introducing ideal elements, viz., infinitely long lines and points at infinity, we can make the theorem that two straight lines always intersect at one and only one point come out universally true. These ideal "infinite" elements have the advantage of making the system of connection laws as simple and perspicuous as possible. Moreover, because of the symmetry between a point and a straight line, there results the very fruitful principle of duality for geometry.. . .

7.2.7.3 George Birkhoff's Axioms for Euclidean Geometry

Although this is not spelled out as such, Birkhoff's geometry[30] is built on a set theory, and it uses **R**, the ordered field of real numbers. His fundamental undefined object is a set of objects called *points*. On this groundwork, he then introduces the *lines* of his geometry; they will be given as sets of points. On top of this, there is a *distance function*; i.e., between any two points A and B: a nonnegative real number $d(A, B)$ such that $d(A, B) = d(B, A)$. And on top of that, one has the structure of *angle*; namely, to any three ordered points $A, O, B(A \neq O, B \neq O)$ there is assigned an $\angle AOB$**; a real number (mod** 2π). The point O is called the **vertex** of the angle.

- **Postulate I**. (Postulate of Line Measure) **The points** $A, B, \ldots$ **of any line** ℓ **can be placed into one-to-one correspondence with the real numbers**, so that for x a non-negative real number, $|xA - xB| = xd(A, B)$ for all points A, B.
- **Postulate II**. (Point-Line Postulate) One and only one line ℓ contains two given points $P, Q \ (P \neq Q)$.
- **Postulate III**. (Postulate of Angle Measure) **The half-lines** $\ell, m \ldots$ **through any point** O **can be put into one-to-one correspondence with the real numbers** a**(mod** 2π**),** so that, if $A \neq O$ and $B \neq O$ are points of ℓ and m, respectively, the difference $am - al$ (mod 2π) is $\angle AOB$. Furthermore if the point B on m varies continuously in a line ℓ not containing the vertex O, the number am varies continuously also.
- **Postulate IV**. (Similarity Postulate) If in two triangles

$$\Delta ABC, \Delta A'B'C'$$

and for some constant $k > 0$,

$$d(A', B') = kd(A, B), \quad d(A', C') = kd(A, C),$$

and

$$\angle B'A'C' = \pm\angle BAC,$$

[30] http://web.mnstate.edu/peil/geometry/C2EuclidNonEuclid/birkhoffs.htm.

then also

$$d(B', C') = kd(B, C), \angle A'B'C' = \pm\angle ABC,$$

and

$$\angle A'C'B' = \pm\angle ACB.$$

Defined Terms
A point B is between A and C $(A \neq C)$, if $d(A, B) + d(B, C) = d(A, C)$. The half-line ℓ' with endpoint O is defined by two points O, A in line ℓ $(A \neq O)$ as the set of all points A' of ℓ such that O is not between A and A'. The points A and C, together with all point B between A and C, for segment AC. If A, B, C are three distinct points, the segments AB, BC, CA are said to form a triangle ΔABC with sides AB, BC, CA and vertices A, B, C.

7.2.7.4 The 'Erlangen Program'

The emergence of non-Euclidean geometries (late 19-th century) clearly required a radical review of the nature of geometry. Felix Klein, a professor in University Erlangen-Nürnberg, proposed the framework of continuous groups as a basis for laying the foundations of geometry.

The Erlangen program takes as a primary geometric notion *the group G of symmetries of the geometry whose underlying space is a set E (homogeneously acted on by G)*. Given only the structure of such a continuous group G, and on fixing a point $p \in E$, given the subgroup $G_p \subset G$ defined by the condition that a symmetry $g : E \to E$ lies in G_p if and only if g fixes p (i.e., $g(p) = p$), one can reconstruct the geometry on E.

For example, for the Euclidean plane its group of symmetries G would be the group of the linear transformations given by compositions of translations and rotations and—taking p to be the origin, say,—G_p could be taken to be the group of rotations about the origin.

(But we might also keeping in reserve the larger groups that include transformations that are flips about a line; and perhaps—for issues of similarity—keeping further in reserve: zooms of the form $(x, y) \mapsto (rx, ry)$).

One can reconstruct all of Euclidean geometry from the group theoretic properties of the inventory just described.

The viewpoint that the *Erlangen Program* takes toward geometry is that starting with a continuous group G (of an appropriate sort) you can construct, and launch, a corresponding geometry where that group G is the group of its symmetries.

7.2.8 From Synthetic to Analytic

The big distinction between the three 'Euclidean axiom-formulations' (Euclid's, Hilbert's, and Birkhoff's) is—I believe—in the implicitly assumed substrates that ground each of the axiom systems:

- Euclid *assumes* that we are—at least vaguely—familiar with the basic nature of 'Euclidean space' and his mission is to describe it more precisely and give terminology so that we may offer reasoned arguments about its features and make constructions in it. *Discuss: proportions versus numbers.*
- Hilbert—in effect—generates his 'Euclidean space' by relational axioms, depending on the substrate (undiscussed explicitly) of set theory.
- Birkhoff brings in (in a way fundamental to his approach) metric considerations; hence his 'substrate' includes quite explicitly the ordered field of real numbers. Moreover, Birkhoff is explicit about the metric of angle-measurement.

The three axiom systems (Euclid's, Hilbert's, Birkhoff's) fall under the general rubric of "synthetic geometries," i.e., set-ups that formulate conditions regarding essentially geometric features. Note that—in contrast to Birkhoff's axioms—there is *no* mention of the *real numbers* or any other number system in Euclid's or Hilbert's Axioms. All three systems are quite different from *'analytic geometry'* which would set things up—from the start—by working in the substrate of $\mathbb{R}^2$ or $\mathbb{R}^3$ and providing geometric definitions in purely algebraic language. Birkhoff's axioms do move closer to that, but are still (interestingly) synthetic.

Here's Felix Klein's definition of the distinction between analytic and synthetic geometry[31]:

> Synthetic geometry is that which studies figures as such, without recourse to formulas, whereas analytic geometry consistently makes use of such formulas as can be written down after the adoption of an appropriate system of coordinates.

and here are his comments in an essay he wrote:

> **On the Antithesis between the Synthetic and the Analytic Method in Modern Geometry[32]:**
>
> The distinction between modern synthesis and modern analytic geometry must no longer be regarded as essential, inasmuch as both subject-matter and methods of reasoning have gradually taken a similar form in both. We choose therefore in the text as common designation of them both the term projective geometry. Although the synthetic method has more to do with space-perception and thereby imparts a rare charm to its first simple developments, the realm of space-perception is nevertheless not closed to the analytic method, and the formulae of analytic geometry can be looked upon as a precise and perspicuous statement of geometrical relations.

[31] Klein, Felix (1948), *Elementary Mathematics from an Advanced Standpoint/Geometry*, New York: Dover (p. 55).

[32] See: Ralf Stephan, *A comparative review of recent researches in geometry* https://arxiv.org/abs/0807.3161.

When I was first learning geometry, Herbert Buseman was the main proponent of keeping as much "synthetic geometry" as possible, but even he realized that he was being out-dated as a purist. He wrote:

> Although reluctantly, geometers must admit that the beauty of synthetic geometry has lost its appeal for the new generation. The reasons are clear: not so long ago synthetic geometry was the only field in which the reasoning proceeded strictly from axioms, whereas this appeal—so fundamental to many mathematically interested people—is now made by many other fields.

There's a good Wikipedia page about this: https://en.wikipedia.org/wiki/Synthetic_geometry.

Given all these sentiments, I take Birkhoff's axioms as being something of a compromise: it is largely synthetic, but with an analytic flavor.

7.2.9 From Axioms to Models: Example of Hyperbolic Geometry

Often—when we frame an axiomatic system—we have a specific structure in mind (Euclid surely did!) and our axiomatic system is a way of allowing us to understand, to study, the structure. Once the axiomatic system is formulated, though, we can reverse the process and ask for concrete structures that *model*[33] (or perhaps are modeled by) our axiomatic system. There are celebrated stories about this issue (*non*-Euclidean geometry) where various slight changes of one postulate ("Euclid's Fifth Postulate") provide axiomatics for various different geometries—Hyperbolic Geometry being one of them.

Essential roles played by models (vis-à-vis axiomatic systems) is that

- the axiomatic systems may elucidate the models;
- the models will establish the *consistency* of the axioms (i.e., proving that they're not self-contradictory);
- the models will offer the intuition needed to think constructively about the axioms.

All this is nicely illustrated by the example of the multiplicity of different models for the same system of axioms of Hyperbolic Geometry. With Hyperbolic Geometry we have an assortment of different *models* any one of which conforms to the axiomatic system of Hyperbolic Geometry.

See, for example: *A brief look at the evolution of modeling hyperbolic space*[34] by Shelby Frazier: *The Mathematics Enthusiast* **14** (2017). This is an excellent topic for our discussion.

---

[33] *Model Theory,* a subject we will touch on later focuses even more explicitly on the relation between setting-up a language-and-structure and models for such.

[34] https://scholarworks.umt.edu/cgi/viewcontent.cgi?article=1387&context=tme.

- For the *Klein model* (used by many) geodesics are straight lines in the (open) unit disc.
- For the *Poincaré disc model* (loved by geometers) geodesics are arcs of circles perpendicular to the boundary. Geodesics are vertical lines from the real axis to infinity; or semicircles perpendicular to the real axis.
- Here's the *Lorentz model*[35] or *hyperboloid model* (loved by physicists).

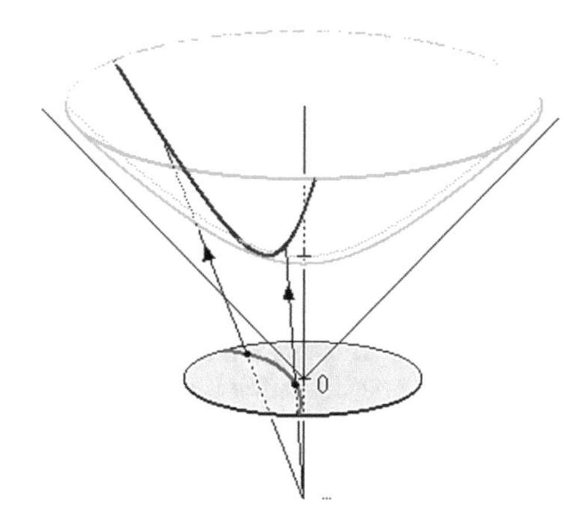

7.3 'Axiomatic Formats' in Philosophy, Formal Logic, and Issues Regarding Foundation(s) of Mathematics and . . . Axioms in Theology

7.3.1 Axioms, Again

It may pay, at this point, to summarize the various uses we have seen of *the axiomatic format*.

(i) *To provide an explicit organized framework for thought:* axioms may be framed to set out the starting assumptions in a line of reasoning. Or, more formally, to formulate a specific algorithmic procedure.
(ii) *To describe in as explicit terms as possible certain specific human interactions:* either fully descriptive (e.g., this *is*—or *may be* a rough model of what people do), or more normative, *desiderata* perhaps. This is as in axiomatic utility theory.
(iii) *To stipulate a 'mathematical structure':* as in axioms for Euclidean or Hyperbolic Geometry. This can be descriptive—i.e., a characterization of a geometry (in the style of Euclid) or more—one might say—ontological (in the style of Hilbert) where the axioms are meant to be an abstract structure that has, as one

[35] https://en.wikipedia.org/wiki/Hyperboloid_model.

of its features, Euclidean Geometry as (a) model. It is a structure that *defines* Euclidean Geometry.

(iv) *To "delineate" a mathematical structure from a previously constructed axiomatic system;* e.g., as in analytic geometry.

7.3.2 Axioms... and 'Psychology'

In axiomatic formulations of models for utility—e.g. axiomatic utilitarianism; or in attempts tò model the manner in which we—individually, or collectively—make choices, the *axioms* play an essential role as the *starting point* of a discussion. Von Neumann and Morgenstern regarded their axioms for expected utility as 'defining' rational behavior in certain risky situations—the axioms reflecting the way that people *should* behave in making rational decisions for their individual or collective choices. The named *paradoxes* (Ellsberg, Allais, St. Petersburg) of Social choice theory make it clear, however, how difficult it is to try to extend rational behavior on the individual level to the social level.

Nevertheless—crudely speaking—*psychology often intervenes,* making these axioms not entirely reliable predictors of behavior, and softening their effect as normative signposts. Kahneman's & Tversky's *Rational choice and the framing of decisions*[36] shows that

- two formulations of the same problem may elicit different preferences, in violation of the axiom of *invariance*; and
- the *dominance rule* is obeyed when its application is transparent, but *dominance* can easily be "masked by a frame in which the inferior option yields a more favorable outcome in an identified state of the world."

> Whether the relation of dominance is detected depends on framing as well as on the sophistication and experience of the decision maker.

In Kahneman & Tversky's language: different *framings* call into play different personal 'takes' on the axioms. A transformation, then, from *rules of rational choice* to the more malleable *rules for subjective reasonable choice*. This was already implicit in Daniel Bernoulli's notion of the concavity of the (function describing) the value of goods, or money. Kahneman & Tversky go much further in considering 'subjective takes on things' (and in many directions: even in their **weight functions** which might be more descriptively labelled as *subjective evaluation of probabilities*).

[36] https://apps.dtic.mil/sti/pdfs/ADA168687.pdf.

7.3.3 Model Theory

Model theory begins by offering a format for doing mathematics within an explicitly shaped 'Language' (in the style of 'universal algebra')—where the 'models' will be *sets with extra structure*—and where its *sentences* interpreted in any 'model' have truth-values that conform to the rules of first-order logic.

The 'opening move' of Model Theory is a powerful and revealing disarticulation of semantics from substance. Here's what I mean: if you are not model-theoretic and want to formulate, say, *graph theory*, you might—for example—just define a **graph** to be given by a set V of vertices and a set E of edges, each edge attaching two distinct vertices and you might also insist that no two vertices are attached by more than one edge. Or you might give a more topological account of this structure; e.g.[37]:

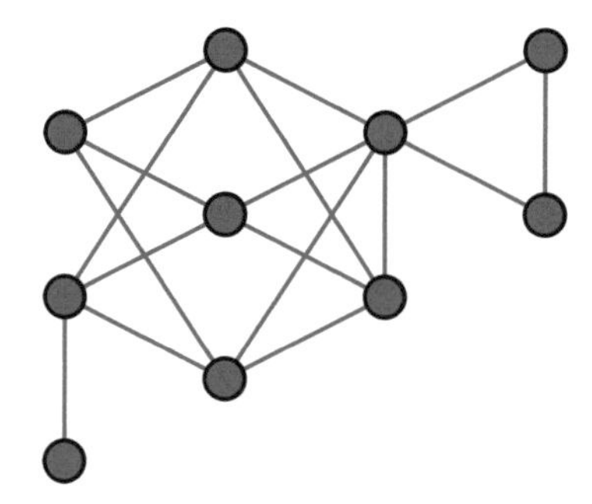

In any event, your formulation begins with a set and then some structure is imposed on it.

Model Theory, reverses this. It begins by offering an explicitly shaped language in which first-order logic is incorporated. In the case of our example of graph theory, the language would have a symbol $\mathcal{E}$ labeled as a *binary relation* (symmetric, but not reflexive) in connection with which we label as *true* sentences:

$$\forall x, y(x \,\mathcal{E}\, y \leftrightarrow y \,\mathcal{E}\, x)$$

and

$$x \,\mathcal{E}\, y \implies x \neq y.$$

An 'interpretation' of this language—or synonymously, a 'model' for this would be a 'representation' of this language in (some version of) Set Theory. That is, it would give us a set V endowed with a binary relation E for which the labeled-as-true sentences are... in fact true; i.e., such a model is simply a graph, where the set of vertices is the set V and the set of edges is given by the binary relation E.

[37] https://en.wikipedia.org/wiki/File:Distance-hereditary_graph.svg.

7.3.4 *Completeness, Consistency*

A Formal System (consisting in, say, a first order theory with a finite collection of axioms) is **consistent** if it is not the case that there exists a well-formed formula P such that P and its negation, $\neg P$, can both be proved from those axioms. It is **complete** if for every well-formed formula P either P or $\neg P$ can be proved.

Take a look at Martin Davis, *Gödel's incompleteness Theorem*, Notices of the AMS **53** (2006) 414–418 http://www.ams.org/notices/200604/fea-davis.pdf.

7.3.5 *Axiomatic Language in Ethics*

There is much to discuss here.[38] The Euclidean format for the organization of rational argument—including first principles of a very general nature (often labelled *Definitions* and *Axioms*) formally set down and referred to explicitly in the justification of each step of arguments that have played important roles in moral philosophy. For example:

7.3.5.1 From: Baruch de Spinoza, *Ethics*

The organization of his *Ethics*[39] has a Euclidean framework where there are very explicitly displayed *Definitions*, *Axioms*, and *Propositions* and where the propositions have arguments labelled *Proof* in which every line of argument refers only to prior *Definitions*, *Axioms*, or *Propositions*.

The long quotation below is from PART II of Spinoza's *Ethics*. I'm eager for some discussion about this in our seminar, since I don't have any idea how Spinoza expects us to deal with this 'borrowed' Euclidean format.

On the Nature and Origin of the Mind

> **DEFINITION I.** By **body** I mean a mode which expresses in a certain determinate manner the *essence of God*, in so far as he is considered as an extended thing. (See Pt. i., Prop. xxv., Coroll.)
>
> **DEFINITION II.** I consider as belonging to **the essence of a thing** that, which being given, the thing is necessarily given also, and, which being removed, the thing is necessarily removed also; in other words, that without which the thing, and which itself without the thing, can neither be nor be conceived.

---

[38] For an overview of some current literature having to do with logic and theology, see the introduction and bibliography of the volume *Logic and the Concept of God* (eds.: Stanislaw Krajewski & Ricardo Sousa Silvestre) Journal of Applied Logics **6** (6): 999–1005 (2019); for the text, go to https://philarchive.org/archive/KRALAT-3.

[39] See https://www.gutenberg.org/files/3800/3800-h/3800-h.htm.

DEFINITION III. By **idea** I mean the *mental conception* which is formed by the mind as a thinking thing.

Explanation: I say conception rather than perception, because the word perception seems to imply that the mind is passive in respect to the object; whereas conception seems to express an activity of the mind.

...

DEFINITION VI. Reality and **perfection** I use as synonymous terms.

...

The definitions of Spinoza vividly connect to (or are contrary to) various anticedent (also succedent) philosophical traditions. Compare, for example, Spinoza's take on the "essence of God" (as related to Definitions I and II above) to the view of St. Thomas Acquinas (as quoted in footnote 2 above). Also note how much is tucked into Definition VI!

AXIOMS

...

AXIOM II. Man thinks.

AXIOM III. Modes of thinking, such as love, desire, or any other of the passions, do not take place, unless there be in the same individual an idea of the thing loved, desired, etc. But the idea can exist without the presence of any other mode of thinking.

...

PROPOSITIONS

PROP. I. Thought is an attribute of God, or God is a thinking thing.

...

Spinoza ends his treatise *Ethics* in a curious manner, offering a quite different take than St. Anselm's proof of the existence of God—or indeed on the ontological proofs offered by Gödel:

PROP. XIII. A mental image is more often vivid, in proportion as it is associated with a greater number of other images.

Proof: In proportion as an image is associated with a greater number of other images, so (II. xviii.) are there more causes whereby it can be aroused. Q.E.D.

PROP. XIV. The mind can bring it about, that all bodily modifications or images of things may be referred to the idea of God.

Proof: There is no modification of the body, whereof the mind may not form some clear and distinct conception (V. iv.); wherefore it can bring it about, that they should all be referred to the idea of God (I. xv.). Q.E.D.

PROP. XV. He who clearly and distinctly understands himself and his emotions loves God, and so much the more in proportion as he more understands himself and his emotions.

Proof: He who clearly and distinctly understands himself and his emotions feels pleasure (III. liii.), and this pleasure is (by the last Prop.) accompanied by the idea of God; therefore (Def. of the Emotions, vi.) such an one loves God, and (for the same reason) so much the more in proportion as he more understands himself and his emotions. Q.E.D.

PROP. XVI. This love towards God must hold the chief place in the mind.

7.3.5.2 The Categorical Imperative

Perhaps the most quoted example of axiomatics in ethics is Immanuel Kant's 'golden rule': the categorical imperative as formulated in his *Critique of Practical Reason*.[40]

> Act in such a way that the maxim of your will can at the same time always hold as a principle of a universal legislation.

This succinct proposition expands impressively in its applications, and is meant to fit into the context of various *Theorems* of morality stated by Kant. Kant insists on the imperative here, noting that "Pure geometry has postulates as practical propositions which, however, contain nothing further than the presupposition that one is *able*" to do something if it were required. Geometry's propositions are, Kant says, "therefore, practical rules under a problematical condition of the will." (What Kant is signaling here is that in Geometry one can construct a circle *or not:* there's no obligation to perform any act—but his theorems of morality oblige 'the will' to act in the above way.)

Earlier in his *Introduction* Kant writes:

> For in the present work we will begin with principles and go to concepts, and only then from these, where possible, continue on to the senses. With speculative reason, in contrast, we began with the senses and had to end with the principles.

which I take to mean that even though he works in a formal setting (with statements labeled as "Theorems," etc.) he is inverting the usual order of appearance of elements in a formal system—starting with conclusions ("principles") and having them reveal basic concepts (an *analysis* of propositional truths rather than a *synthesis*).

7.3.5.3 Ontological Arguments

But the most curious engagement of axiomatics in ethics are the various "Ontological arguments" related to the existence (or at least to the definition) of God. These arguments can be essentially pro- (i.e., claiming that God's existence is proved) as in St. Anselm, or Spinoza; or essentially critical as in Aquinas or Kant. An enlightening account of these arguments, and their history, can be found in the Stanford Encyclopedia of Philosophy's entry https://plato.stanford.edu/entries/ontological-arguments/. (Read especially Sects. 1–3.)

A curious common thread in many of the ontological arguments is to allow 'existence' to be a possible predicate (or not!) of the various things-of-thought. St. Anselm, for example, puts a value judgment on this predicate: *it's more perfect to exist than not!*.[41] Compare this with Spinoza's Definition VI quoted above.

---

[40] See p. 38 (**Fundamental Principle of Pure Practical Reason**) in http://www.kantwesley.com/Kant/CritiqueOfPracticalReason.pdf.

[41] *Old joke:*
A: I wish I never was born!
B: Oh, only one in a million is that lucky.

So if you conjure the most perfect thing-of-thought that can be conceived, well: if it doesn't exist, there's your contradiction. For now imagine whatever it is that you conjured up, but as *existing*, and you've just conceived of a yet more perfect thing-of-thought—voilà.

Often in these ontological arguments one sees the *unqualified* use of the quantifier $\exists$ to establish *existence* (of something) as being a predicate (of that something). That is: one asserts existence of an entity, without specifying in what realm that entity is (so-to-speak) 'taken from.'

In symbols: as long as you have a set in mind as your domain of discourse—call it Ω—it makes sense to consider formulas such as:

$$\exists x \in \Omega \text{ such that } \dots,$$

but you're asking for trouble if you have no specific set such as Ω in mind and just want to deal with the formula:

$$\exists x \text{ such that } \dots.$$

(This puts such arguments in the same framework as *unqualified* use of the quantifier $\forall$, as is behind Russell's paradox and the various uses of 'unqualified universal quantification', related to the classical *crisis in the foundations of mathematics*.)

Baruch de Spinoza, however, in his *Ethics* has—as far as I can make out—a very different take. Spinoza gives three different 'proofs'. These might be characterized as follows.

(i) Two versions of the 'ontological argument':

- God's essence (simply) entails existence.
- The potentiality of non-existence is a negation of power, and contrariwise the potentiality of existence is a power, as is obvious.

(ii) A version of the principle of insufficient reason:

> If, then, no cause or reason can be given, which prevents the existence of God, or which destroys his existence, we must certainly conclude that he necessarily does exist. If such a reason or cause should be given, it must either be drawn from the very nature of God, or be external to him—that is, drawn from another substance of another nature. For if it were of the same nature, God, by that very fact, would be admitted to exist. But substance of another nature could have nothing in common with God (by Prop. ii.), and therefore would be unable either to cause or to destroy his existence.

7.3.6 *Axiom-Definitions in Classical Physics*

We have already encountered instances of 'organizing statements' setting up an axiomatic framework, where it isn't quite clear to what extent a statement is a *defini-*

tion of something or an *axiom* about the behavior of something.[42] A simple example of this is the labelled Definition (xvi) in Euclid's *Elements*: *And the point is called the centre of the circle.* That this point is 'unique' has 'gone without saying,' i.e., the definition carries along with it the uniqueness assertion, as axiom.

Newtons three Laws form–or at least suggest—a formal axiom system. Specifically interesting for our seminar is the structure, the nature, and various historical revisions imposed on these axioms. Moreover, at times these laws move from their status as axiom (the status they enjoy in straight Newtonian mechanics) to *part-definition* or to something less easily describable. Here are Newton's Laws, not in their original language, but in their original strength and intent as they were given in Newton's 1687 treatise: *Mathematical Principles of Natural Philosophy*:

- **First law:** In an inertial frame of reference, an object either remains at rest or continues to move at a constant velocity, unless acted upon by a force.
- **Second law**: In an inertial reference frame, the vector sum of the forces F on an object is equal to the mass m of that object multiplied by the acceleration a of the object:

$$F = ma.$$

- **Third law**: When one body exerts a force on a second body, the second body simultaneously exerts a force equal in magnitude and opposite in direction on the first body.

These are the starting laws, that launched that extraordinarily exact science—classical mechanics.

7.3.6.1 Kant's Metaphysical Counterpart to Newton's Laws

Almost exactly a century later (1786) Kant published a treatise *Metaphysical Foundations of Natural Science*[43] as an attempt to revisit Newton's laws from a metaphysical perspective—and in a wildly idiosyncratic way. In contrast to Newton's very often-quoted "Hypotheses non fingo," (I make no—metaphysical—hypotheses) Kant is not shy of making them! For example, here's Kant's definition of **matter**:

> **Matter** is whatever is movable and can be an object of experience.

The first definition in his first chapter is:

> **Definition 1:** I call something 'material' if and only if it is movable in space. Any space that is movable is what we call 'material' or **relative space**. What we think of as the space in which all motion occurs—space that is therefore absolutely immovable—is called 'pure' space or **absolute space**.

---

[42] A historically (and substantially) important example of a shift from definition to law, and perhaps back again, occurs in the early discussions of what it means to be computable. See "Not a Definition, a Natural Law" in Allyn Jackson's essay *Emil Post: Psychological Fidelity* published in the journal *Inference* https://inference-review.com/article/psychological-fidelity.

[43] http://www.earlymoderntexts.com/assets/pdfs/kant1786.pdf.

Starting with this definition—to express it anachronistically—Kant is working out the issue of dependence on frame of reference—in particular the idea that motion is a relative notion. He proclaims as a *Principle:*

> Every motion that could be an object of experience can be viewed either as
>
> - the motion of a body in a space that is at rest or as
> - the rest of a body in a space that is moving in the opposite direction with equal speed. It's a free choice.

(This might be a point for discussion in our seminar, since, here and at other places in his essay, Kant is formulating something that—in modern language—might be phrased as the question of whether a *coordinate-free language* is possible for the framing of physical laws.)

Regarding gravitational attraction, he offers this description, but labelled as propositions:

> **Proposition 7:** The attraction that is essential to all matter is an unmediated action through empty space of one portion of matter on another.
>
> **Proposition 8:** The basic attractive force, on which the very possibility of matter depends, reaches out directly from every part of the universe to every other part, to infinity.

And eventually, Kant formulates his version of the laws:

- **First law of mechanics:** Through all changes of corporeal Nature, the over-all amount of matter remains the same—neither increased nor lessened.
- **Second law of mechanics:** Every change in matter has an external cause. (Every motionless body remains at rest, and every moving body continues to move in the same direction at the same speed, unless an external cause compels it to change.)
- **Third mechanical law:** In all communication of motion, action and reaction are always equal to one another.

(A further possible topic for discussion might be whether Kant succeeds or fails in giving Newton's laws a metaphysical underpinning—or in elucidating the underlying issues. Or: what is Kant actually trying to get at?)

7.3.6.2 Ernst Mach's Reconfiguration of Newton's Laws

Going forward yet another century, there is Ernst Mach's retake.[44] Most striking is his view of Newton's second Law. Simply formulated above, it was that the *law* $F = ma$ might be viewed as defining the concept of mass: the quantity m is, in fact *defined by its appearance as a constant* in such a law. Specifically, this second law is the following assertion:

> Appropriately understood —relative to any specific body (conceived of as "point-mass") the 'ratio' F/a is constant; and this constant is *defined* to be the mass of this body.

[44] *The Science of Mechanics* (transl.: T. J. McCormack) Open Court Publications. Co, (1919) https://archive.org/details/scienceofmechani005860mbp/page/n5/mode/2up.

This turns the tables, a bit, on the second law, focusing on its role as a definition of one of the three components of this Newtonian system. Indeed Mach had a few variant formulations of this definition: one of them in terms of a little drama involving a closed system consisting of two bodies A and B doing whatever they are doing to each other (repelling or attracting) and as a result having accelerations a_A and a_B at which point (making implicit use of Newton's *Third law*) one can *define the ratio of the masses* of bodies A and B to be:

$$m_A/m_B = -a_B/a_A,$$

which has the added virtue of not ever dealing explicitly with the notion of Force.[45]

7.3.6.3 D'Alembert's Revision

More than a century earlier than Mach, there appeared a different rewriting of Newton's second law known (now) as D'Alembert's Principle.[46] This principle seems to have been created by a shockingly simple—you might think:—trick. But, in fact, it represents an extremely important change of viewpoint related to (and perhaps inspiring) an entire genre of *conservation laws* and *stationary principles*. Here is D'Alembert's simple idea. Write Newton's second law, $F = ma$, in this (clearly equivalent) way:

$$F - ma \quad = \quad 0.$$

Dubbing "$-ma$" as a sort of fictional force (referred to as *inertial force*) you get that the equilibrium of the system is marked (in this somewhat semantic juggle) as the sum of all the forces on the system being: 0. A *conservation* of forces. This is the precursor of two other rewritings of Newton's laws—these going under the names of *Lagrange*, and *Hamilton*. (An interesting theme for a final paper would be a discussion of the way in which these formulations differ from each other, and from Newton's original formulation.) Viewing Newton's second law (as D'Alembert did) as a 'conservation law' is in the spirit of other conservation laws, such as conservation of energy—which in a sense was more of a principle to be defended rather than a pure axiom, in that whenever, in some set-up, the 'conservation principle of energy' seemed to be violated, the physicists confronted by this seeming violation fashioned a new facet or form of energy to put into the equation so as to maintain conservation.

[45] For a very critical discussion of the merits of Mach's approach, see *There Is No Really Good Definition of Mass* by Eugene Hecht http://physicsland.com/Physics10_files/Mass.pdf.

See also *About the definition of mass in (Machian) Classical Mechanics* by Marco Guerra and Antonio Sparzani (Foundations of Physics Letters, **7**, No. 1, 1994)].

[46] See the discussion beginning p. 340 in Ernst Mach's *The Science of Mechanics* https://archive.org/details/scienceofmechani005860mbp/page/n363/mode/2up.

7.4 Listable Sets, Gödel's Incompleteness Theorem, and Algorithms

7.4.1 *Readings*

(i) Martin Davis, *Gödel's incompleteness Theorem*, Notices of the AMS **53** (2006) 414–418. http://www.ams.org/notices/200604/fea-davis.pdf.
(ii) Allyn Jackson, *Emil Post: Psychological fidelity*, Inference, **4** (2018). https://inference-review.com/article/psychological-fidelity.
(iii) Wikipedia on *Gödel's incompleteness Theorem*. https://en.wikipedia.org/wiki/G%C3%B6del%27s_incompleteness_theorems.

7.4.2 *How Rigorous Will We Be?*

The answer is: not rigorous at all, and in no way self-contained; I shall only hint at the train of arguments—introducing a minimum of notation—but I'll try to usefully evoke the general sense of these arguments, and be as honest as I can be, in all this 'evoking.'

7.4.3 *Listable Sets of Integers*

(synonyms: *recursively enumerable, computably enumerable, algorithmically enumerable*).

Let's start with some examples of sets that are easy to "list"

•
$$2, 3, 5, 7, 11, 13, 17, 19, 23, \ldots$$

•
$$2!, 3!, 4!, 5!, \ldots$$

Discuss what is meant by *easy*.

Generally, a subset $\mathcal{L} \subset \mathbb{Z}$ is called **listable**[47] if there exists a finite computer program whose output gives a sequence $\alpha_1, \alpha_2, \alpha_3, \ldots$ of integers such that the set $\mathcal{L}$ is precisely this collection of numbers; i.e.,

[47] A very readable introduction to a bit of this theory (and especially the historical and personal context in which it arose) is Allyn Jackson's essay; see D(ii) above. For a pretty readable intro to the notion of *listable set* see https://en.wikipedia.org/wiki/Recursively_enumerable_set noting that listable is called **recursively enumerable** there. For a hint about hierarchies of 'relative listability' take a quick look at https://en.wikipedia.org/wiki/Post%27s_theorem.

$$\mathcal{L} = \{\alpha_1, \alpha_2, \alpha_3 \ldots\}.$$

A computer algorithm that does this job will be called a computer algorithm that "lists $\mathcal{L}$".

Note, though, that–even if the computer spits out a "new" integer every second—the ordering in which the numbers in the computer's listing of $\mathcal{L}$ come may be very up and down in terms of size. Therefore if you suspect that a given number, say 3, is *not* in $\mathcal{L}$ and need to have a definite guarantee of the truth of your suspicion, well (if you are right!) running the computer algorithm for any finite length of time—with the helter-skelter sizes of numbers that come up will be of no help to you: if 3 does show up, it is—of course–therefore in $\mathcal{L}$; if it hasn't yet shown up, no matter how long the computer has run, this tells you nothing about whether it is or isn't in $\mathcal{L}$.

For example, (I'm taking an offhand random example) consider the set of numbers that are expressible as a sum of two sixth powers minus a sum of two sixth powers. Now the set $\mathcal{L}_o$ of such numbers is listable. There is a simple way of systematically listing all such numbers. Run systematically through all quadruples of whole numbers A, B, C, D organizing these quadruples by size in what is called 'diagonal ordering' and collecting the values $n := A^6 + B^6 - C^6 - D^6$, this constituting a list of the elements of $\mathcal{L}_o$.

What are the numbers n that are *not* in $\mathcal{L}_o$? I (personally) don't know[48] whether, say, 8 is or is not in $\mathcal{L}_o$.

Suppose you have a listable subset of positive numbers:

$$\mathcal{L} \subset \mathbf{N} := \{1, 2, 3, \ldots\}.$$

By the **complement** of such a set $\mathcal{L}$ one means the set of all numbers in $\mathbf{N}$ that are *not in* $\mathcal{L}$. Call this complement $\mathcal{L}^{\perp}$. So the union of $\mathcal{L}$ and $\mathcal{L}^{\perp}$ is all of $\mathbf{N}$:

$$\mathcal{L} \cup \mathcal{L}^{\perp} = \mathbf{N}.$$

Lemma 7.1 Day and Night: *There is an algorithm to determine whether any number* $n = 1, 2, 3, \ldots$ *is contained in* $\mathcal{L}$ **if and only if** *both* $\mathcal{L}$ *and* $\mathcal{L}^{\perp}$ *are listable.*[49]

Proof If there is such an algorithm, go through the positive integers one by one, and for each integer n use the algorithm to determine whether it is or isn't contained in $\mathcal{L}$ and put that number n in the appropriate listing of $\mathcal{L}$ if the algorithm says it is, and in $\mathcal{L}^{\perp}$ if the algorithm says it isn't.

Going the other way, suppose that both $\mathcal{L}$ and $\mathcal{L}^{\perp}$ are listable. For any integer n spend your days listing $\mathcal{L}$ and your nights listing $\mathcal{L}^{\perp}$ and you are guaranteed to find n at some day or night; this gives us the algorithm that determines whether n is or isn't contained in $\mathcal{L}$.

---

[48] ... but I haven't thought much about how difficult it is to know....

[49] In the literature, if a set $\mathcal{L}$ has the property that both $\mathcal{L}$ and $\mathcal{L}^{\perp}$ are listable, then $\mathcal{L}$ is called **recursive**.

7.4.4 Emil Post's Fundamental Discovery

A fundamental result of Cantor is that there are sets that are unlistable. This follows from the fact that the set of all *listable* sets is a countable set of sets because the set of all possible lists is countable... but the set of *all* sets is an uncountable set. So there has to exist some unlistable set.

But a fundamental discovery of Emil Post is that there exists a listable set $\mathcal{L}$ whose complement $\mathcal{L}^{\perp}$ is *unlistable*.[50] It follows (e.g., from Lemma 7.1) that there is no decision procedure to determine whether any given integer n is or is not contained in $\mathcal{L}$.

Much of what comes later are afterthoughts to—and elaborations of—this discovery!

7.4.5 Gödel's Incompleteness Theorem

Suppose we have a formal system $\mathcal{F}$ that is **consistent**—i.e., is such that for no wff P is it true that P and $\neg P$ is provable in $\mathcal{F}$—and has a rich enough vocabulary to perform whatever it is required to do below.[51] We aren't being at all explicit here, so take this as a (perhaps overly relaxed) way of directly getting to the heart of the idea of the incompleteness theorem.

For our formal system $\mathcal{F}$ there is a language, we have the standard apparatus and clear rules of inference etc., so we can actually algorithmically list all well-formed-formulae. Do that for $m = 1, 2, 3, \ldots$, giving us a complete list

$$m \mapsto P_m \tag{7.2}$$

that runs through every well-formed formula.

Definition 7.1 One says that a formal system is **complete** if for *any* proposition P formulated in the language of the system either the proposition P or its negation $\neg P$ is provable.

Now take *any one* of Post's sets

$$\{1, 2, 3, \ldots\} \leftrightarrow \mathcal{L} \tag{7.3}$$

[50] I.e., following the previous footnote: (in the literature) the set $\mathcal{L}$ would be called *recursively enumerable* but not *recursive*. For a generously detailed, and beautiful account of this theorem, and the general properties of the concepts *recursively enumerable* and *recursive* see Emil Post's *Recursively enumerable set of positive integers and their decision problems*: https://projecteuclid.org/download/pdf_1/euclid.bams/1183505800. And see p. 291 specifically.

[51] For example: we are requiring our formal systems to be **effectively axiomatized** (also called effectively generated) so that its set of theorems is a listable set. That is, there is a computer program that, in principle, could enumerate precisely the theorems of the system. The standard formal systems Peano arithmetic and Zermelo-Fraenkel set theory (ZFC) have that property.

listable in the language of $\mathcal{F}$ such that the complement $\mathcal{L}^{\perp}$ is unlistable.

Theorem 7.4 *(Gödel) There is at least one positive integer ν for which neither the statement:*

$$P: \ \nu \ is \ not \ in \ \mathcal{L}$$

nor its negation

$$\neg P: \ \nu \ is \ in \ \mathcal{L}$$

is provable in the formal system $\mathcal{F}$.

Proof Euclidean geometry without the parallel postulate is incomplete, because some statements in the language (such as the parallel postulate itself) can not be proved from the remaining axioms.

Proof Working systematically, take each positive integer n and spend your days and nights this way. For a fixed n here is the procedure, which I'll call **Proc**(n):

> **Each Day:** look at successive 100 entries in the list (7.3) of elements of $\mathcal{L}$ to check if n is among those entries. If, on some day, you find n in $\mathcal{L}$ your work is done. Just remember all that work.
>
> **Each Night:** examine successive 100 well-formed formulae in the list (7.2) to see if, for some P_m among those entries, P_m *is a proof that* ν *is not in* $\mathcal{L}$. If, on some night, you find that one of those well-form formulae P_m is a proof that ν is not in $\mathcal{L}$ your work is done. Just remember it.

OK, here are the possibilities:

(i) The 'procedure' **Proc**(n) terminates finitely for *every* positive integer n. That is, we have an algorithm that for all n determines in finite time whether n is or is not in $\mathcal{L}$.
(ii) There "is" a positive integer ν such that the 'procedure' **Proc**(ν) never terminates.

Consider, first, (i) above: if (i) held it would give us a finite algorithm to list the elements of $\mathcal{L}^{\perp}$: for every n run systematically through your days and nights, throwing out the n's that show up in daytime but list the n's for which a proof P_m (that $n \in \mathcal{L}^{\perp}$) has been found in those nighttimes. This, in effect, gives an algorithmic listing of $\mathcal{L}^{\perp}$—*contrary to assumptions*. So, (i) cannot occur.

This leaves (ii). Note that such a number ν cannot be a member of $\mathcal{L}$, for if it were, it would be found some day. So

- ν *is* a member of $\mathcal{L}^{\perp}$; i.e., the statement ν *is in* $\mathcal{L}$ is not provable; and:
- there's also no proof—in the formal system $\mathcal{F}$—of the negation of this—i.e., of the statement ν *is* **not** *in* $\mathcal{L}$—since if there were, the procedure **Proc**(ν) would terminate. QED

Questions

(i) Why did I put quotation-marks in the "is" in the statement of (ii) above?
(ii) How does this discussion change when I change the formal system $\mathcal{F}$ (within which we are working)?

7.4.6 A Diophantine (Synonym: 'Arithmetic') Formulation: The Result of Matiyasevich-Robinson-Davis-Putnam: A Counter-Statement to Hilbert's Tenth Problem

Theorem 7.5 *(MRDP) For any listable set of positive integers $\mathcal{L}$ there is a polynomial $p(t, x_1, x_2, \ldots, x_d)$ in some finite number $(d + 1)$ of variables with integer coefficients such that n is in $\mathcal{L}$ if and only if there are integers $(a_1, a_2, \ldots, a_d)$ such that*

$$p(n, a_1, a_2, \ldots, a_d) = 0. \tag{7.4}$$

Corollary 7.7 *(Unsolvability: a counter-statement to Hilbert's Tenth Problem) There is a polynomial $p(t, x_1, x_2, \ldots, x_d)$ with integer coefficients for which there is no algorithm to determine for n running through all positive integers whether $p(n, x_1, x_2, \ldots, x_d)$ has a solution (i.e., integers $(a_1, a_2, \ldots, a_d)$ satisfying Eq. 7.4).*

Proof of Corollary 7.7 (given Theorem 7.5): Just take any listable set $\mathcal{L}$ for which $\mathcal{L}^{\perp}$ is unlistable, use Theorem 7.5 to find the corresponding polynomial, and interpret what this means.

If our students wish, we may go a bit into these topics:

(i) Various ways of thinking about Gödel's Incompleteness Theorem. E.g., "independence."
(ii) Peano Arithmetic, ZFC.
(iii) Gödel's view of Set Theory.
(iv) Current 'programs': Harvey Friedman; Hugh Woodin.
(v) ... but we will at least ask—and briefly discuss—the question:
what is a set?

7.4.7 What Is a Set?

I'm not sure we have a definitive answer to this yet. A 'set' is a pretty lean mathematical object, evoked—if not captured—by the simple phrase *a collection of things*. Nevertheless *sets* provide the substrate for such a wide variety of mathematical objects. So, an axiom system that 'models' set theory is clearly of foundational importance in mathematics.

That the axiom system for Set Theory is free of contradictions (i.e., is 'consistent') is, of course, necessary; and we should take particular care to achieve consistency, especially with Russell's paradox as a cautionary tale.

The most common such axiom system, *Zermelo-Fraenkel Set theory* referred to as ZF—or if one adds *the axiom of choice* to its list of axioms, one calls it ZFC—was proposed in the early twentieth century by Ernst Zermelo and Abraham Fraenkel.

Instead of presenting this axiom system as a 'formulated thing,' it might be more engaging if we discuss it, building it (or something close to it) up, by stages in conversation. The most dramatic procedure for such building-up is due to Von Neumann.

We allow our *set theory* to have the standard first-order logic as semantics so we can use, for example, the standard quantifications $\forall$ and $\exists$, etc.

Since we are trying to capture the notion of *set* as a formal entity having the intuitive meaning of *collection of things*, the most economical thing one might do—and ZF does this!—is to

- have just one type of object in our vocabulary (these objects are either to be thought of as *sets* or *objects* that are members of some set or of some sets; or these objects are both sets in themselves and members of other sets: all this depends on their properties in connection with the unique relation ($\in$) that will be introduced in the bullet below; these objects will be designated by some letter (e.g., $x, y, \ldots$)

  and
- have just one formal relation (*membership*, denoted $\in$). So, for x, y in our discourse, it might be the case that $x \in y$—namely, x is a member of the set y. It might also be the case that $x \in y \in z$ (i.e., y is a member of the set z but is also a set in its own right, containing x as a member).

7.4.7.1 The Ambiguity of "and so on." Building Sets; as von Neumann Did

Start with the set containing no element, the *empty set* { } which we'll call $\varnothing$. We then—ridiculous as this may seem—consider the set containing only one element, the empty set: $\{\varnothing\}$. Well then, we can imagine keeping going: form the set containing the two elements:

$$\{\ \varnothing, \{\varnothing\}\ \},$$

the set containing the three elements:

$$\{\ \varnothing, \{\varnothing\}\ \{\ \varnothing, \{\varnothing\}\ \}\},$$

and so on...

Namely, this procedure proposes to construct an unlimited collection of sets (from nothing). In fact even the elements of these sets are, curiously, themselves sets. So *all the objects* of this discourse launched by von Neumann are themselves sets (!) Can we produce a system of axioms that

- formalizes the construction that von Neumann has proposed (so every object is a set) and
- provides a formal architecture reasonable for Set Theory?

The axiomatic system ZF consists of eight axioms: and if one throws in the very tricky Axiom of Choice as ninth axiom one calls the system ZFC.

7.4.7.2 Equality and 'Extensionality'

Discuss "extension versus intension"
We want two objects a, b of our discourse to be regarded as *equal* if—in English—every element of the set a is equal to an element of the set b and vice versa; *and also* every set containing a as an element also contains b as an element and vice versa. If you are worried about this looking suspiciously like a circular definition, in that the word *equal* appears in the formulation—how about:

$$\forall z[z \in x \Leftrightarrow z \in y] \wedge \forall w[x \in w \Leftrightarrow y \in w].$$

Now the **Axiom of Extensionality** can be formulated as

$$\forall x \forall y[\forall z(z \in x \Leftrightarrow z \in y) \Rightarrow \forall w(x \in w \Leftrightarrow y \in w)],$$

which says that if x and y have the same elements, then they belong to the same sets; i.e., they're equal.

Axioms that produce sets from other sets

7.4.7.3 Definition

A set z is a **subset** of a set x if and only if every element of z is also an element of x:

$$(z \subseteq x) \Leftrightarrow (\forall q(q \in z \Rightarrow q \in x)).$$

7.4.7.4 The Axiom of Specification: An Axiom Guaranteeing That We May Create a Subset of a Given Set z by Imposing Conditions on the Elements of z

If z is a set and $\phi(x)$ is some formula imposing a condition on the variable x then

$$A := \{x \in z : \phi(x)\}$$

is a set too.

Note that a formula $\phi(x)$ *alone* is not enough to produce a set (one of the troublesome features of various ontological proofs of the existence of God): this axiom requires you to stipulate a set z from which you want to cut out a piece (as subset) by imposing some predicate as a condition.

7.4.7.5 The Axiom of Power Sets: An Axiom Saying (Essentially) That the 'Collection' of all Subsets of a Given Set Is Again Itself a Set (in Its Own Right)

The Axiom of Power Set specifically states that for any set x, there is a set y that contains every subset of x:

$$\forall x \exists y \forall z [z \subseteq x \Rightarrow z \in y].$$

The axiom schema of specification can then be used to define the power set P(x) as the subset of such a y containing the subsets of x exactly:

$$P(x) = \{z \in y : z \subseteq x\}.$$

7.4.7.6 Axiom of Pairing

If x and y are sets, then there exists a set B which contains x and y as elements.

$$\forall x \forall y \exists B (x \in B \wedge y \in B).$$

(The axiom schema of specification can be used to reduce this to a set with exactly these two elements.) Note that taking $x = y$ we then get that if x is any set, the singleton $\{x\}$ (set containing x as its unique member) is again a set.

7.4.7.7 Axiom of Union

The union over the elements of a set exists.

The axiom of union states that for any *set of sets* $\mathcal{F}$ there is a set C containing every element that is a member of some member of $\mathcal{F}$.

Letting $S(w)$ abbreviate $w \cup \{w\}$ where w is some set, we get—from the Axioms of pairing and union that $S(w)$ is indeed a set. So, for example, assuming that von

Neumann's $\varnothing$ is assumed to exist as a set in our theory, each of the following creatures in von Neumann's list also are sets in our theory:

$$\varnothing, \quad \{\, \varnothing, \{\varnothing\}\, \}, \quad \{\varnothing, \{\varnothing\}\, \{\, \varnothing, \{\varnothing\}\, \}\}, \quad \text{etc.}$$

Note that it would be perfectly consistent with the collection of axioms discussed so far to have a model of set theory where every set is finite. In fact, there might be no sets at all in our model, so ...

7.4.7.8 Axiom of Infinity

Recall that $S(w) := w \cup \{w\}$ where w is a set. The Axiom of Infinity states that there exists a set X such that the empty set $\varnothing$ is a member of X and, whenever a set y is a member of X, then $S(y)$ is also a member of X. In effect, we are requiring that the ('infinite') union of the sets in von Neumann's list is *also a set in our theory.*

$$\exists X\, [\varnothing \in X \wedge \forall y (y \in X \Rightarrow S(y) \in X)]\,.$$

Now it would be good to know that X has infinitely many members (this issue is hinted at in the scare-quotes around the word "infinite" above). We need that the sets listed in the definition of X are all different, because if two of these sets are the same, the sequence will loop around in a finite cycle of sets. A Russellian quandary.

The *axiom of regularity* below is a clever way of establishing a basic property of the relation $\in$ and preventing loops from happening:

7.4.7.9 Axiom of Regularity

We include as axiom the requirement that every non-empty set x contains a member y such that x and y are disjoint sets.

$$\forall x\, (x \neq \varnothing \rightarrow \exists y \in x\, (y \cap x = \varnothing)).$$

The remaining one (or two) axioms

The above evocation of the first 7 of the 8 axioms of ZF is meant to spark a discussion—I hope there's time for such discussion in our last sessions. We have not discussed the 8-th axiom **The Axiom of Replacement** which asserts that images of (appropriately) definable functions from sets to set, $f : S \rightarrow T$, are again sets; nor the final axiom **The Axiom of Choice** (that turns ZF to ZFC). But I hope we can talk a bit about the impact of Gödel's Incompleteness on the project of setting up a formal system that is comprehensive enough to be a 'foundation' for mathematical practice. And on the connection with higher cardinalities. And... the important issue of mathematical induction. And... of course... so on.

Barry Mazur has been at Harvard University for over sixty years. He first came there as a Junior Fellow (at the Society of Fellows) in 1959 and is now the Gerhard Gade University Professor at Harvard. His initial interests in mathematics were in Topology, then Dynamical Systems, and then Algebraic Geometry; his current work is largely in Number Theory. He has given seminar-courses (jointly with other professors) in the History of Science Department, the Philosophy Department and at Harvard's Law School. His essay in this volume consists of notes for the students of such a seminar-course ("Axiomatic Reasoning"), reflecting on the possible directions such a seminar might take.

Chapter 8
Axiomatic Thinking, Identity of Proofs and the Quest for an Intensional Proof-Theoretic Semantics

Peter Schroeder-Heister

Abstract Starting from Hilbert's *Axiomatic Thinking*, the problem of identity of proofs and its significance is discussed in an elementary proof-theoretic setting. Identifying two proofs, one of which is obtained from the other one by removing redundancies, leads, when used as a *universal* method, to a collapse of all proofs of a provable proposition into one single proof and thus trivialises proof identity. Principles of proof-theoretic harmony with *restricted* means of redundancy reduction might be used instead, though this limits one to a certain format of formal systems. The discussion of proof identity suggests the claim that annotations of proofs telling the reader which rule is applied at a particular step, must be considered part of the proof itself. As a general perspective, it is advocated that the investigation of intensional aspects of proofs should be given more space in proof theory and proof-theoretic semantics.

8.1 Introduction: *Axiomatic Thinking* and Hilbert's Programme

In his essay *Axiomatic Thinking* David Hilbert argues that it is necessary "to study the essence of mathematical proof itself if one wishes to answer such questions as the one about decidability in a finite number of operations" [12, p. 414 (orig.), p. 1115 (transl.)]. "We must [. . .] make the concept of specifically mathematical proof itself into an object of investigation" (415/1115). Taking his later conception of 'proof theory' into account and what has afterwards been called "Hilbert's programme", this can be read as the claim that we should conceive proofs as formal proofs within a formal system, of which we can then, by manipulating them as formal objects, hopefully demonstrate that they never generate formal contradictions. A strong method in proving such consistency is to reduce the system in question to other systems whose consistency has already been established. In fact, the significance of such a reductive approach is already emphasised in *Axiomatic Thinking* by mentioning the potential

P. Schroeder-Heister (✉)
Department of Computer Science, University of Tübingen, Sand 13, 72076 Tübingen, Germany
e-mail: psh@uni-tuebingen.de

F. Ferreira et al. (eds.), *Axiomatic Thinking I*,
https://doi.org/10.1007/978-3-030-77657-2_8

reduction of arithmetic and set theory to logic [12, p. 412 (orig.), p. 1113 (transl.)]. In this sense *Axiomatic Thinking* can be read as the starting point of 'reductive' [24] proof theory whose programme is to establish advanced systems as conservative (and thus consistent) extensions of more elementary systems. This is incorporated in the finitist programme of justifying infinitist ways of reasoning as extensions of finite ways of reasoning such that at least the consistency of these systems can be proved by finite means, even if they are not conservative extensions [13].

As is well known, the original form of Hilbert's programme failed due to Gödel's second incompleteness theorem, according to which the inference methods codified in an elemementary system such as Peano arithmetic do not suffice, for reasons of principle, to demonstrate the consistency of the system in question. Hilbert's programme nevertheless initiated the development of mathematical proof theory which investigates, among many other issues, the strengh of formal systems and their relative reducibility, the expressive power of such systems, including what can be reached by various forms of induction principles incorporated in such systems. As regards consistency proofs, Gerhard Gentzen's work [11] constituted the first pioneering achievements, above all his consistency proofs for arithmetic using transfinite induction.

8.2 *Axiomatic Thinking*, General Proof Theory and Proof-Theoretic Semantics

In parallel to the development of proof theory in the spirit of Hilbert's programme, so-called *general proof theory* took ground. General proof theory is interested in proofs as fundamental entities being used in deductive sciences. Here the problem of consistency, which was the starting point of Hilbert's programme, is not in the centre of interest. Of course, consistency is essential for proofs. But it is simply not the leading point of view from which proofs are looked at. One could say that in general proof theory we are not primarily interested in the *result* of proofs, that is, in the assertions that are proved or can be proved in a proof system, but in the *form* of proofs as representing arguments. Philosophically speaking, general proof theory deals with *intensional* aspects of proofs, while proof theory in the spirit of Hilbert's programme, which is interested in the logical power of proof systems, with their *extensional* aspects.

In fact, the initial quotation above from *Axiomatic Thinking* shows that the programme of general proof theory is present already in Hilbert. He explicitly speaks of the "essence" of proofs and the "concept of [...] proof itself", which is exactly what general proof theory is all about. And in the paragraphs ahead of this quote he discusses in detail the problem of entirely different methods of proving the same geometric claim [12, pp. 413–414 (orig.), pp. 1114–1115 (transl.)], which means that the idea of different proofs of a mathematical theorem and thus the problem of the identity and difference of proofs is on his agenda. In this sense it would be wrong to claim that general proof theory and the interest in proofs in themselves is something totally different from what Hilbert had in mind when creating his proof theory. Even though consistency-oriented proof theory strongly dominated Hilbert's later writings, general proof theory was always in the background, and in *Axiomatic Thinking* still on equal level with reductive proof theory.

Note that the view of proofs as *formal proofs*, that is, as proofs in a formal system, versus proofs as *arguments*, that is, as entities conveying epistemic force, is not the dividing line between reductive or consistency-oriented and general proof theory. It is certainly true that when studying consistency or the reduction of theories, we are studying syntactic properties of proofs,[1] while when considering proofs as arguments, we are studying epistemic and semantic properties going beyond the syntactic level. However, even in the second case, we are still considering formal proofs, as these epistemic and semantic properties are *properties of formal proofs*, namely as formal proofs being representations of arguments. So ontologically it is the same sort of entities which are discussed in reductive and general proof theory. This is analogous to the situation we have in model theory, where we look at syntactically specified formulas and theories from a semantic perspective (in the sense of a denotational semantics).

The interdependency of consistency-oriented proof theory and general proof theory is fully clear in the work of Gentzen, who is at the same time the exponent of proving consistency and of laying the grounds for general proof theory. The latter is due to the fact that in his seminal *Investigations into logical deduction* [10] Gentzen created the *calculus of natural deduction* as formal system that is very near to actual reasoning, in particular to reasoning in mathematical practice. In the same work he developed the *calculus of sequents*, which is very well suited for certain proof-theoretic investigations. Gentzen's formal systems as well as the results he obtained for his systems are highly significant both for reductive and general proof theory. This holds in particular for his method of cut elimination for sequent systems, which is fundamental for reductive proof theory and likewise for general proof theory.

The term "general proof theory" as well as its explicit proclamation as a research programme is due to Dag Prawitz [24, 26], after he had already, in his 1965 monograph *Natural Deduction*, provided the first systematic investigation of Gentzen's calculus of natural deduction [23]. At the same time and with a similar target, Georg Kreisel [16] had proposed a modification of Hilbert's programme towards the study of proofs as genuine objects, and not only as tools for the investigation of derivability and consequence. On the philosophical side, Michael Dummett [7] was outlining his programme of a verificationistic theory of meaning, which took place in parallel to Prawitz's notion of proof-theoretic validity [27, 29]. Roughly at the same time Per Martin-Löf's type theory emerged [19, 33], which built on closely related logical foundations, and which laid new foundation of mathematics as an alternative to set theory and to Frege's and Russell's type-theoretic conceptions.

[1] As far as Hilbert is concerned, this holds at least with respect to his later conception of proof theory, but, as pointed out by a reviewer of this paper, perhaps not yet at the stage of *Axiomatic Thinking*, where it is not fully clear that the consistency problem should be solved syntactically.

For these and related approaches the author proposed the term "proof-theoretic semantics" [30, 31]. The reason for choosing this term was to emphasise that such investigations belong to philosophical semantics, and that therefore the term "semantics" should not be left to denotational semantics alone. Philosophically, general proof theory and proof-theoretic semantics belongs to the theory of *meaning as use* [39] and more specifically, to an inferentialist theory of meaning [2], though with many additional inspirations from ideas and results in proof theory [22].

General proof theory is a proof theory based on philosophical interests. This does not mean that no mathematical methods can enter when these interests are pursued. On the contrary, the application of mathematical methods on syntactically coded proofs delivers basic philosophical insights. These insights concern in particular the problem of the *identity of proofs*, which is the main topic of this paper. Identity of proofs is not currently the central theme of general proof theory. However, it should be in the centre of interest, because it is immediately connected to the question of the essence of proofs.

In fact, in his proclamation of general proof theory, Prawitz pointed out that one of the basic topics of this discipline is the identity of proofs, namely the question, when syntactically different proofs of the same theorem should be considered 'essentially' identical and when they should be considered 'essentially' different [24, p. 237].[2] This coincides with the discussion of Hilbert in *Axiomatic Thinking* about different proofs for the same result. Hilbert's emphasis was on *conceptually* different proofs in the sense that these proofs used different methods or even came from different branches of mathematics, that is, proofs using different proof ideas.

We are still far from being able to formally elucidate what a proof idea is. However, as a first step we will discuss at the level of natural deduction proofs, using its very elementary *conjunctive fragment*, what identity of proofs can mean and which problems are connected with it. This is at least in the spirit of what Hilbert meant by "making the concept of specifically mathematical proof itself into an object of investigation" and what Prawitz had in mind when putting forward the idea of a general proof theory. What we are going to tell will be aporetic in many respects. Even in the context of our tiny fragment of natural deduction considerable problems show up. However, we hope to convince the reader of the fundamental fact that there is something on the intensional side of proofs, in addition to what is being proved, something that Hilbert called the "essence" of proofs. As a prominent example, we discuss the *redundancy criterion* for proof identity, according to which proofs are identical, when they only differ by adding or removing redundancies, and point to problems associated with this criterion. As an important side product, we conclude that the *annotations of proofs* should be considered ingredients of the proofs themselves. That is, the explicit specification which step we want to apply at a certain place, especially if the shape of the step leaves this open, is more than a metalinguis-

[2] Prawitz does not use the term "essentially", but speaks of "synonymy" of derivations and "equivalence" of proofs. See also Prawitz [25, p. 132] and Kreisel's discussion [16, p. 127] of this problem, Feferman's [9] review of Prawitz [24], and Došen's introduction to the symposium on General Proof Theory at the 14th Congress of Logic, Methodology and Philosophy of Science (Nancy 2011) [5].

tic comment on a proof, but belongs to the proof itself. This we see as an indication that the notion of *intension* is related to the notion of *intention* even in the area of formal reasoning.

8.3 Identity of Proofs

Quine's slogan "no entity without identity" is one of the cornerstones of ontological reasoning in the philosophy of language [28]. It is based on the claim that, in order to refer to an individual entity, we need a criterion that tells us of (purported) entities a and b whether they are different, or whether they are perhaps a single entity referred to in different ways. If we apply this idea to mathematical proofs, this means that a mathematical proof can only be individuated as an entity, if we have a criterion that tells us of *syntactically different* proofs $\mathcal{D}$ and $\mathcal{D}'$ whether, *with respect to their content*, they should be considered the same proof or not.

Quite independently of the philosophical problem of individuation, according to which without an identity criterion we cannot speak of an individual entity, it is simply mathematically interesting to know whether two proofs, which prima facie look different, are nevertheless 'essentially' the same proof. Working mathematicians often have quite strong intuitions about whether two proofs of the same theorem are based on the same proof idea, and they often agree with respect to these intuitions.

As the concept of *proof idea* is not capable of a precise rendering, at least not with the current conceptual tools of mathematical or philosophical logic, we confine ourselves to extremely simple proofs which are formulated in a very small fragment of elementary logic. More precisely, we consider formal proofs which are only formulated by means of logical conjunction $\wedge$. By that we mean proofs in which only the conjunctive composition of sentences is made explicit. For such proofs, we have three proof rules: One introduction rule and two elimination rules.

The introduction rule

$$\frac{A \quad B}{A \wedge B} \wedge I \tag{8.1}$$

allows us to generate, from proofs $\begin{matrix}\mathcal{D}_1 \\ A\end{matrix}$ of A and $\begin{matrix}\mathcal{D}_2 \\ B\end{matrix}$ of B, a proof

$$\frac{\begin{matrix}\mathcal{D}_1 \\ A\end{matrix} \quad \begin{matrix}\mathcal{D}_2 \\ B\end{matrix}}{A \wedge B} \wedge I$$

of $A \wedge B$. The expression to the right of the inference line denotes the rule being applied ("I" for "introduction"). The elimination rules for conjunction are

$$\frac{A \wedge B}{A} \wedge E_1 \qquad \frac{A \wedge B}{B} \wedge E_2 \tag{8.2}$$

They allow us to recover, from a proof $\genfrac{}{}{0pt}{}{\mathcal{D}}{A \wedge B}$ of $A \wedge B$, proofs

$$\genfrac{}{}{0pt}{}{\mathcal{D}}{\dfrac{A \wedge B}{A} \wedge E_1} \qquad\qquad \genfrac{}{}{0pt}{}{\mathcal{D}}{\dfrac{A \wedge B}{B} \wedge E_2}$$

of A and of B. As we will see below, it is important to distinguish the two $\wedge$-elimination rules ("E" for "elimination") terminologically by an index ("1" and "2", respectively). The rule with index 1 picks the left argument of conjunction, the rule with index 2 the right one. Mathematically, we can consider the introduction rule for conjunction as the formation of a *pair* of proofs, and the elimination rules as the *projections* of such a pair on its left or right component. This very elementary framework of conjunction logic is already sufficient to point to basic problems, results and difficulties in connection with the problem of identity of proofs.

There are two opposite extremes in answering the question concerning the identity of proofs, which are equally inappropriate and both trivialise the idea of identity. One extreme consists in considering proofs $\mathcal{D}$ and $\mathcal{D}'$ to be identical, if they are identical as syntactic objects. This criterion is too narrow, as any syntactic modification of a proof, so tiny and minor it may be, would result in a different proof, though the 'content' of the proof has not changed at all. Two syntactically different proofs of a proposition A could never be identical. The other extreme consists in in considering proofs $\mathcal{D}$ and $\mathcal{D}'$ to be identical, if they are proofs of the same proposition A. This criterion is too wide. Because all syntactically different proofs of a provable proposition A could be identified, every provable proposition A would have only one single proof. In fact, in many areas we are solely interested in whether a proposition A is provable or not—for example whether in a theory a contradiction "C and not-C" is provable. Whether there are potentially different proofs of a proposition would then be irrelevant. However, in general proof theory we pursue the idea that the study of proofs goes beyond the study of provability. This means that in principle, though perhaps not in every single case, there can be different proofs of a provable proposition A.

Thus we need to define a plausible equivalence relation on the class of syntactically specified proofs of a proposition A, which is neither *syntactic identity* (every syntactic proof of A constitutes a singleton equivalence class) nor the universal relation (all syntactic proofs of A belong to the same equivalence class). If $\mathcal{D}$ and $\mathcal{D}'$ are proofs of A, we would like to define a *nontrivial* equivalence relation $\mathcal{D} = \mathcal{D}'$, which comes as near as possible to our intuitive idea that $\mathcal{D}$ and $\mathcal{D}'$ represent *the same* proof of A.

As to our terminology: When we talk of the identity of proofs $\mathcal{D}$ and $\mathcal{D}'$, and express this as $\mathcal{D} = \mathcal{D}'$ by means of the identity sign "=", then we always mean the equivalence relation *to be explicated*. When we talk of the syntactic identity of proofs, we always say this explicitly, but never use the identity sign for it. If $\mathcal{D}$ is a proof of A, we often write $\genfrac{}{}{0pt}{}{\mathcal{D}}{A}$. The expression $\genfrac{}{}{0pt}{}{\mathcal{D}}{A}$ then denotes the same as $\mathcal{D}$—the A below $\mathcal{D}$ only serves to mention the proposition being proved and is not an extension of $\mathcal{D}$.

8.3.1 The Redundancy Criterion

One possibility to define identity between proofs is to point out certain redundancies in proofs and to specify procedures removing these redundancies. A proof $\mathcal{D}$ would then have to be considered identical to a proof $\mathcal{D}'$, if $\mathcal{D}'$ results from $\mathcal{D}$ by such a removal of redundancies. In natural deduction, a prominent case of that kind is the introduction of a proposition immediately followed by its elimination. This situation can be clarified by analogy with arithmetical operations.

In algebra we are often dealing with structures where with a given operation an inverse operation is associated, such as in the case of groups. If, for example, we add the integer b to an integer a and immediately afterwards subtract it, we obtain the very same integer a back:

$$(a + b) - b = a$$

At the level of proofs in natural deduction we have a similar situation, as the elimination rules are inverses of the introduction rules. In the fragment considered here, the calculus for conjunction, these are the rules (8.1) and (8.2).

Analogously to the example of addition and substraction, introduction and elimination rules cancel each other out. Consider the introduction of a conjunction followed by its elimination, passing from given proofs $\begin{matrix}\mathcal{D}_1\\A\end{matrix}$ and $\begin{matrix}\mathcal{D}_2\\B\end{matrix}$ for A and B by $\wedge$-introduction to their conjunction and going back to A by $\wedge$-elimination:

$$\dfrac{\dfrac{\begin{matrix}\mathcal{D}_1\\A\end{matrix}\quad\begin{matrix}\mathcal{D}_2\\B\end{matrix}}{A \wedge B}\,\wedge I}{A}\,\wedge E_1 \tag{8.3}$$

Then this is obviously a redundancy, since we had already proved the proposition A before engaging in these two inference steps, namely as the left premiss of the first step. According to the redundancy criterion we want to identify two proofs, one of which is nothing but a redundant form of the other one. Therefore we postulate the following identity:

$$\dfrac{\dfrac{\begin{matrix}\mathcal{D}_1\\A\end{matrix}\quad\begin{matrix}\mathcal{D}_2\\B\end{matrix}}{A \wedge B}\,\wedge I}{A}\,\wedge E_1 \quad = \quad \begin{matrix}\mathcal{D}_1\\A\end{matrix} \tag{8.4}$$

Correspondingly we postulate the following identity, in which the first projection is replaced with the second projection:

$$\dfrac{\dfrac{\begin{matrix}\mathcal{D}_1\\A\end{matrix}\quad\begin{matrix}\mathcal{D}_2\\B\end{matrix}}{A \wedge B}\,\wedge I}{B}\,\wedge E_2 \quad = \quad \begin{matrix}\mathcal{D}_2\\B\end{matrix} \tag{8.5}$$

In accordance with Prawitz [23] such identities are also called "reductions", as they reduce the redundancy in a proof. We also speak of "redundancy reductions". Since in the theory of natural deducion, (8.4) and (8.5) are always postulated, we call these identities *standard reductions* for conjunction (later we will consider a further standard reduction). Corresponding standard reductions can be given for all other logical signs and also for non-logical operations.

These reductions can also be formulated algebraically, if we consider proof rules as functions I, E_1, E_2 transforming given proofs into new proofs. Then the $\wedge$-introduction rule generates from two proofs $\mathcal{D}_1$ and $\mathcal{D}_2$ for A and B, respectively, a new proof $I(\mathcal{D}_1, \mathcal{D}_2)$ of their conjunction, and the elimination rules generate from a proof $\mathcal{D}$ of a conjunction $A \wedge B$ proofs $E_1(\mathcal{D})$ and $E_2(\mathcal{D})$ of A and B, respectively. The standard reductions (8.4) and (8.5) then become the identitities

$$E_1(I(\mathcal{D}_1, \mathcal{D}_2)) = \mathcal{D}_1 \qquad E_2(I(\mathcal{D}_1, \mathcal{D}_2)) = \mathcal{D}_2 \tag{8.6}$$

The theory of natural deduction based on standard reductions was developed by Dag Prawitz in his groundbreaking monograph *Natural Deduction* [23], as was the idea of defining the identity of natural deduction proofs by reference to these reductions: "Two derivations represent the same proof if and only if they are equivalent" [24, p. 257], where equivalence is established by applying standard reduction steps.[3] Being redundancy reductions, the standard reductions can be generalised as follows.

Obviously, the standard reductions follow the following general pattern:

$$\frac{\begin{array}{c}\mathcal{D}\\ A\end{array}}{\begin{array}{c}\vdots\\ \hline A\end{array}} \;=\; \begin{array}{c}\mathcal{D}\\ A\end{array} \tag{8.7}$$

For the case of (8.4) $\begin{array}{c}\mathcal{D}\\ A\end{array}$ corresponds to the proof $\begin{array}{c}\mathcal{D}_1\\ A\end{array}$, and for the case of (8.5) the proposition A corresponds to B and the proof $\begin{array}{c}\mathcal{D}\\ A\end{array}$ corresponds to $\begin{array}{c}\mathcal{D}_2\\ B\end{array}$. All other parts of these proofs are represented in (8.7) by dots.

The idea behind (8.7) is the following: We disregard the potential proof steps between the upper and the lower A and only focus on the situation, in which we start with a proof $\begin{array}{c}\mathcal{D}\\ A\end{array}$ of A and then return to A in a way not further specified. As the steps leading from $\begin{array}{c}\mathcal{D}\\ A\end{array}$ to A are redundant, we can identify the extended proof of A with the initial proof of A. We call this identification the *general redundancy reduction*. As just explained, the standard reductions (8.4) and (8.5) are two particular cases of it, in which a *specific* form of redundancy, namely introduction immediately followed by

[3] More precisely, equivalence is defined as the transitive and reflexive closure of reducibility by standard reductions.

elimination, is considered. What we call "general redundancy reduction" is discussed in Ekman [8].

Unfortunately, the general redundancy reduction has unwanted consequences. Consider the following situation, in which we consider any two given proofs $\mathcal{D}_1$ and $\mathcal{D}_2$ of a proposition A, used in the following extended proof of A:

$$\dfrac{\dfrac{\begin{matrix}\mathcal{D}_1 & \mathcal{D}_2 \\ A & A\end{matrix}}{A \wedge A}}{A} \tag{8.8}$$

Here it does not make a difference, of whether $\wedge$-elimination is conceived as left or right projection. In (8.8), there are obviously two possibilities to apply the general reduncy reduction. If we identify the lower A with the left upper A, we obtain the identity

$$\dfrac{\dfrac{\begin{matrix}\mathcal{D}_1 & \mathcal{D}_2 \\ A & A\end{matrix}}{A \wedge A}}{A} = \begin{matrix}\mathcal{D}_1 \\ A\end{matrix} \tag{8.9}$$

If we identify the lower A with the right upper A, we obtain

$$\dfrac{\dfrac{\begin{matrix}\mathcal{D}_1 & \mathcal{D}_2 \\ A & A\end{matrix}}{A \wedge A}}{A} = \begin{matrix}\mathcal{D}_2 \\ A\end{matrix} \tag{8.10}$$

The identities (8.9) and (8.10) immediately give us

$$\begin{matrix}\mathcal{D}_1 \\ A\end{matrix} = \begin{matrix}\mathcal{D}_2 \\ A\end{matrix}$$

Since $\mathcal{D}_1$ and $\mathcal{D}_2$ are arbitrary proofs of A, the general redundancy reduction allows the identification of arbitrary proofs of the same proposition A.

In this way the identity of proofs becomes the universal relation, which means that the equivalence relation of identity is trivialised in one of the two ways discussed above. Thus the redundancy criterion for identity fails. Note that this result does not depend on how exactly the standard reductions are formulated. Depending on whether the elimination step in (8.8) is conceived as left or right projection, either (8.9) or (8.10) is a standard reduction in the sense above. However, this fact plays no role as the standard reductions are instances of the general redundancy reduction. Therefore, if we assume the introduction and elimination rules (8.1) and (8.2) as rules governing conjunction, then the general redundancy reduction trivialises the identity of proofs (see also [32]).

8.3.2 An Example from Mathematics

Considering conjunctions of the form $A \wedge A$ may appear artificial. To supersede this objection we consider as a concrete example Euclid's theorem according to which there are infinitely many prime numbers, and denote it by P_∞. Furthermore, we consider two proofs of this theorem that rely on completely different concepts, for example the number-theoretic proof by Euclid himself, here denoted as $\mathcal{D}_{Euclid}$, and the proof by Euler which uses elementary calculus, here denoted as $\mathcal{D}_{Euler}$ (see, e.g., [1]). If we combine these two proofs conjunctively,

$$\dfrac{\begin{array}{c}\mathcal{D}_{Euclid}\\ P_\infty\end{array} \qquad \begin{array}{c}\mathcal{D}_{Euler}\\ P_\infty\end{array}}{P_\infty \wedge P_\infty}$$

we have a duplication of the theorem P_∞, but keep at the same time the information both of Euclid's and of Euler's proof. From the two proofs we are forming a pair of proofs that comprises both. No information contained in any of these proofs is lost.

From this pair of proofs we can recover the respective proof by means of right or left projection: By left projection Euclid's proof

$$\dfrac{\dfrac{\begin{array}{c}\mathcal{D}_{Euclid}\\ P_\infty\end{array} \qquad \begin{array}{c}\mathcal{D}_{Euler}\\ P_\infty\end{array}}{P_\infty \wedge P_\infty}}{P_\infty}\wedge E_1 \quad = \quad \begin{array}{c}\mathcal{D}_{Euclid}\\ P_\infty\end{array}$$

and by right projection Euler's proof

$$\dfrac{\dfrac{\begin{array}{c}\mathcal{D}_{Euclid}\\ P_\infty\end{array} \qquad \begin{array}{c}\mathcal{D}_{Euler}\\ P_\infty\end{array}}{P_\infty \wedge P_\infty}}{P_\infty}\wedge E_2 \quad = \quad \begin{array}{c}\mathcal{D}_{Euler}\\ P_\infty\end{array}$$

The kind of projection (left or right) tells us which proof we get back. We can, of course, ignore the kind of projection and thus waive proof information. That is, we can consider the proof

$$\dfrac{\dfrac{\begin{array}{c}\mathcal{D}_{Euclid}\\ P_\infty\end{array} \qquad \begin{array}{c}\mathcal{D}_{Euler}\\ P_\infty\end{array}}{P_\infty \wedge P_\infty}}{P_\infty} \tag{8.11}$$

simply as a structure leading to P_∞, *in whatever way* we have obtained the conjunction standing above P_∞. Nothing speaks against this way of proceeding. We must only be content with the fact that the proof achieved proves the same, namely P_∞, but that neither the proof information from $\mathcal{D}_{Euclid}$ nor that from $\mathcal{D}_{Euler}$ is available any more, after we refrained from labelling the last step of (8.11) either as left or as right projection. Based on our proof, we continue to have the right to assert P_∞, because our proof ends with this proposition. However, we can neither identify this proof with $\mathcal{D}_{Euclid}$ nor with $\mathcal{D}_{Euler}$, which was still possible, when the step to P_∞ was considered a projection. In the case of (8.11), we have, so to speak, in the course of the deviation via $P_\infty \wedge P_\infty$, thrown away our 'luggage' in form of proof information, even though the legitimacy of the claim P_∞ is not affected. By means of the deviation we have not simply created redundancy in the sense of additional unnecessary information, but conversely destroyed the information which would allow us to identify the proof reached with one of the proofs we started with.

8.3.3 Harmony Instead of Reduction of Redundancy

The standard reductions alone do not trivialise the identity of proofs, as can be seen relatively easily.[4] This suggests to refer, in the definition of identity of proofs, only to the standard reductions, rather than considering the general redundancy reduction. This requires the philosophical task to elucidate what is the distinguishing characteristics of the standard reductions beyond the fact that they are cases of the general redundancy reduction, that is, that they reduce redundancy in proofs.

Here the concept of *harmony* comes into play, by means of which the relationship between introduction and elimination rules is frequently characterised (the term goes back to Dummett, see [31, 37]). Consider again the case of conjunction, where we have the situation that the *conditions of the introduction* match with the *consequences of the eliminations*. The condition of the introduction of $A \wedge B$ is the pair consisting of A and B, and the consequences of the eliminations are again this pair obtained by left and right projection.

If according to (8.3) one moves from the conditions of the introduction rule to the consequences of the elimination rules, by first introducing a conjunction and immediately afterwards eliminating it, then this is a step from a proposition A to its harmonious counterpart, that is, from a part of the condition of the introduction rules to a part of the conclusion of the elimination rules. That one does not gain any new information, is not only due to the fact that in both cases we deal with the proposition A, but because one is using the complementary steps of introduction and elimination rules, which cancel each other out due to the harmony between these rules.

In this way we even obtain a sequence of steps dual to the one considered. That the conditions of the introduction rules match the consequences of the elimination rules also means that one does not lose anything when applying an elimination rule. This means that from applications of the eliminations rules to $A \wedge B$, that is, from the consequences of $A \wedge B$, we can, by means of the introduction rules, go back to $A \wedge B$. This corresponds to the reduction

[4] By using the uniqueness of normal forms for identical proofs (Church–Rosser); see, e.g., Hindley [14].

$$\frac{\dfrac{\begin{array}{c}\mathcal{D}\\ A \wedge B\end{array}}{A}\wedge E_1 \qquad \dfrac{\begin{array}{c}\mathcal{D}\\ A \wedge B\end{array}}{B}\wedge E_2}{A \wedge B}\wedge I \quad = \quad \begin{array}{c}\mathcal{D}\\ A \wedge B\end{array} \tag{8.12}$$

which is here also considered a standard reduction.[5] Algebraically this corresponds to the equation

$$I(E_1(\mathcal{D}), E_2(\mathcal{D})) = \mathcal{D} \tag{8.13}$$

Therefore the idea behind this approach is that due to the matching of the conditions of introduction with the consequences of elimination one has pairs of completely symmetric inference steps, which represent a *specific* form of redundancy reduction. The standard reductions for conjunction (including (8.12)) express the complementarity of the steps introduction-elimination or elimination-introduction, and it is this specific form of redundancy reduction, which makes the standard reductions non-trivial. This is opposed to general redundancy reduction (8.7), where between the occurrences of A, which are identified, there may lie a non-specified proof section rather than just a pair of complementary rule applications. It is possible to show that the standard reductions (8.4), (8.5), (8.12) are maximal in the sense that no further identitites may be postulated without trivialising the notion of identity [6]. This maximality result is often considered the distinguishing feature of the standard reductions, turning them into a proper base for proof identity.

Of course, it is not a philosophical necessity to base the notion of identity of proofs on the notion of harmony, that is, on symmetries between introduction and elimination rules for logical signs. Even the maximality result just mentioned does not force us to that conclusion. It cannot be excluded that there are different postulated sets of identities, which are likewise maximal. However, currently the harmony principle appears to be the only plausible way to motivate the standard reductions as sensibly restricting the general redundancy reduction, which, as we have seen, goes too far.

For further discussion of the identity of proofs from the logical and philosophical point of view see [3, 4]. For harmony in relation to identity of proofs see [37, 38].

8.3.4 The Annotation of Proofs

When carrying out a proof, one justifies one's steps by telling *which* inference step one is just performing. In our simple case of conjunction we have written the designation of the rule used next to the inference line. Very frequently one finds the opinion that these annotations are nothing but metalinguistic comments that only serve to *explicate* what one is doing, without adding anything to the proof step itself.

From the point of view of identity of proofs this view is misguided, at least in its general form. In most cases it is obvious which rule has been applied in an inference step, simply because, due to the syntactic form of the propositions involved, only

[5] Reading (8.12) from right to left, Prawitz [24] speaks of "expansions".

one single rule fits to the step. For the $\wedge$-introduction rule this is always the case, because a constellation

$$\frac{\begin{array}{cc}\mathcal{D}_1 & \mathcal{D}_2 \\ A & B\end{array}}{A \wedge B}$$

must be an application of $\wedge$-introduction, whatever form A and B have. For the elimination rules this is not always the case. If we apply an elimination rule to the proposition $A \wedge A$:

$$\frac{\begin{array}{c}\mathcal{D} \\ A \wedge A\end{array}}{A}$$

then, since the right and left component of $A \wedge A$ are identical, this can be an application of the left projection $\wedge E_1$ as well as of the right projection $\wedge E_2$. To disambiguate the situation, we write either

$$\frac{\begin{array}{c}\mathcal{D} \\ A \wedge A\end{array}}{A} \wedge E_1$$

or

$$\frac{\begin{array}{c}\mathcal{D} \\ A \wedge A\end{array}}{A} \wedge E_2$$

This means that the annotation ("$\wedge E_1$" or "$\wedge E_2$") is part of the proof, as it gives information needed to understand it. We cannot refrain from deciding between $\wedge E_1$ und $\wedge E_2$. Otherwise we would have to accept both

$$\frac{\dfrac{\begin{array}{cc}\mathcal{D}_1 & \mathcal{D}_2 \\ A & A\end{array}}{A \wedge A}}{A} = \begin{array}{c}\mathcal{D}_1 \\ A\end{array}$$

and

$$\frac{\dfrac{\begin{array}{cc}\mathcal{D}_1 & \mathcal{D}_2 \\ A & A\end{array}}{A \wedge A}}{A} = \begin{array}{c}\mathcal{D}_2 \\ A\end{array}$$

as valid identities, and therefore the identification or arbitrary proofs $\mathcal{D}_1$ and $\mathcal{D}_2$ of A. This was exactly the situation found with the general redundancy reduction, in which it played no role which elimination rule was applied.

After realising that the annotation of the rule being used is part of the proof itself, we can modify the notion of proof by turning the annotion into what is proved. The proof step

$$\frac{A \wedge A}{A} \wedge E_1$$

would have to be written as

$$\frac{A \wedge A}{E_1 : A}$$

As the premiss $A \wedge A$ would be annotated itself with an annotation t, one would write:

$$\frac{t : A \wedge A}{E_1(t) : A}$$

Since in this way an annotation contains all annotations of steps above, the annotation of a proven proposition codes the proof of this proposition. Thus the necessity to consider the annotations of proof steps as parts of the proof, leads to the idea to associate with a proven proposition the coding of its proof. That is, what is actually proven is not the proposition A, but the judgement (claim) $t : A$, where t stands for the proof itself.

Here we can bring to bear the functional view of proof steps mentioned in Sect. 8.3.1 and consider the annotation $E_1(t)$ to be a function applied to t, and postulate certain equations corresponding to the standard reductions. In our case these are the equations (8.6) and (8.13).

This leads to the basic idea of constructive type theories, since the judgement $t : A$, which is now the 'proper' claim in a proof, in contradistinction to the proposition A alone, is structurally related to the assertion that an object t has the type A. This relationship cannot be discussed here. It underlies in particular Martin-Löf's type theory, which in the recent two decades has gained strong ground in general mathematics through Vojvodsky's homotopy theoretic interpretation [33, 34]. The motivation for this conception is normally quite different from what we have presented here. Our philosophical motivation was that annotations of proofs belong to the claims to be proved, so that codes of proofs become a natural ingredient of what is proved.

The idea that the 'proper' structure of a proved proposition A is $t : A$, where t is the code of the proof of A, is often viewed as an argument for or against the identification of certain proofs. However, this is only partially conclusive. The standard reductions cannot be justified that way. The situation (8.3), in which elimination follows to introduction, would now be displayed as

$$\frac{\dfrac{\begin{array}{cc}\mathcal{D}_1 & \mathcal{D}_2 \\ t_1 : A & t_2 : B\end{array}}{I(t_1, t_2) : A \wedge B}}{E_1(I(t_1, t_2)) : A}$$

In order to identify $t_1 : A$ and $E_1(I(t_1, t_2)) : A$, we would need to presuppose the identity

$$E(I(t_1, t_2)) = t_1$$

and thus one of the identities (8.6), which are motivated by the standard reductions. However, even though we cannot obtain a justification of the standard reductions, we obtain a refutation of the general redundancy reduction (8.7). This general reduction would require that in the situation

$$\dfrac{\dfrac{\begin{array}{c}\mathcal{D}\\ t : A\end{array}}{\vdots}}{t' : A}$$

the judgements $t : A$ and $t' : A$ can be identified, which is not possible, if there is no reason to assume $t = t'$. Such a reason is not available for unspecified t and t'. The universal assumption $t = t'$ expresses that arbitrary proofs of A can be identified, which is the trivialisation of proof identity which we do not intend. Therefore, if we accept the idea of proof annotations as parts of proofs, we have an argument against the general redundancy reduction, which unlike the argument in Sect. 8.3.1 does not refer to the reductions themselves, but only to the identifiability of assertions. As far as the justification of the standard reductions as basis of proof identity is concerned, the harmony of introduction and elimination rules continues to be the starting point. A suitable annotation and decoration discipline certainly helps to avoid *unwanted* identifications of proofs, but does not by itself provide the *intended* identification of proofs inherent in harmony principles.

8.4 Conclusion

Why is our result not satisfactory in every respect? For the very restricted context of conjunction logic we have shown that general redundancy is not suitable as an identity criterion for proofs, as it leads to a trivialisation of the concept of proof identity. As the logic of conjunction is normally available in any logical system, this result has wide consequences. The non-suitability of the general redundancy criterion holds for virtually any system.

It would have been the advantage of the general redundancy criterion that its formulation is independent of which logical framework is used, and which rules of proof are available. To admit redundancy reduction only in connection with the harmony of introduction and elimination rules, means a very significant restriction: This identity criterion is only available for proof systems, which build exclusively on harmonic rules. This is the case in (constructive) propositional and predicate logic. Even in constructive type theories one tries to carry these harmony principles through all the rules. But is it necessary that proof systems are always structured that way? Already bivalent classical logic falls out of this framework. Does this mean that it does not make sense to speak of proof identity in classical logic? Should it not be possible to develop proper and non-trivial proof identity criteria for logics which are *not* based on a constructive conception of proof-theoretic harmony? Answering these questions seems to us to be a central desideratum of a proof-theoretic semantics of non-constructive logics.

The aporetic character of our considerations also shows how far proof theory is still away from the treatment of 'proper' proofs in mathematics, and how it is even further away from the explication of the 'idea' behind a given mathematical proof. It is proof ideas in which mathematicians are basically interested, when they compare proofs, as Hilbert does in *Axiomatic Thinking*. Mathematical and philosophical proof theory is only slowly progressing towards this problem.[6] On the other hand, we must concede to proof theory that it has developed a precise syntactic concept of 'proof' which allows one to formulate in it concrete mathematical proofs. Recent proof-theoretic research on the foundations of mathematical concept formation and reasoning show an increased colloboration between philosophy, mathematical logic and mathematical practice which gives hope for progress.[7]

The discussion of identity of proofs demonstrates that intensional considerations in proof-theoretic semantics are needed, if we are not only interested in what can be proved in a given system, but also, and perhaps primarily, in what a proof is and how we carry out proofs. *Proof theory* that deserves its name, should be more than a tool in a *theory of provability*.

What we left out of the picture drawn here is the relation between proofs and algorithms,[8] which is very narrow, in particular with respect to the decoration of proofs by means of certain annotations. As these annotations can be viewed as proof terms, they can motivate a functional view of proof reductions and identity along the lines of the Curry-Howard-correspondence. However, while this would essentially be a re-iteration of the discussion of redundancy reductions by using term equations, it might be more interesting to compare intensional proof theory with the algorithmic view of intensions in the spirit of Moschovakis [21], by relating his abstract concept of an algorithm to the functional concepts used in type-theoretic proofs. This would also allow us to link to the debate about intensions in natural language semantics, which is the field where the problem of intensions showed up first, and where it is still most prominent.

Our plea for an intensional proof-theoretic semantics may be rounded up with a short philosophical remark on the notion of 'intention', which in the philosophical tradition has sometimes been related to the notion of 'intension' (e.g., Hintikka, [15]). We have argued that the annotations of a proof belong to the proof itself by giving the example of a step of conjunction elimination where, without annotation, the left and right projection are not distinguishable. The fact that such a step is to be considered, for example, as a right projection, can be viewed as *my intention* when carrying out the proof. That this step is a right projection, is how I want this step to be understood when

[6] *Mutatis mutandis* this holds for the even more advanced question of wether two proofs of a proposition are not just identical or non-identical, but whether one is simpler than the other, which is the content of Hilbert's 24th problem [35, 36].

[7] See, for example, the activities of the *Association for the Philosophy of Mathematical Practice* (www.philmathpractice.org).

[8] As remarked by one of the reviewers of this paper.

giving my argument, and that this understanding can play a significant proof-theoretic role is precisely what we tried to show in the previous section. This demonstrates that 'deep' philosophical questions of the theory of intentions and actions are not far from general considerations concerning the meaning of proofs. Semantics and action are interrelated even in logic.[9]

Acknowledgements I am grateful for the invitation to participate in the conference "Axiomatic Thinking" in Zurich in September 2017. Some of the ideas presented there have also been presented in November 2017 at the annual fall meeting of the Mathematical Society in Hamburg and published (in German) in the *Mitteilungen der Mathematischen Gesellschaft in Hamburg*, 38 (2018), 117–137. I thank the participants of both conferences for stimulating discussions. I would also like to thank Luca Tranchini for the critical discussion of an earlier version of the manuscript and two anonymous reviewers for helpful comments and suggestions. The final version of this paper was written during a stay at the Swedish Collegium for Advanced Study (SCAS) in Uppsala, and I am very grateful to the Collegium for their hospitality and support.

References

1. Aigner, M. and Ziegler, G. M. (2010). *Proofs from THE BOOK*. Springer, Berlin, 4th edition.
2. Brandom, R. B. (2000). *Articulating Reasons: An Introduction to Inferentialism*. Harvard University Press, Cambridge Mass.
3. de Castro Alves, T. (2018). *Synonymy and Identity of Proofs: A Philosophical Essay*. Dissertation Universität Tübingen, Tübingen.
4. Došen, K. (2003). Identity of proofs based on normalization and generality. *Bulletin of Symbolic Logic*, 9:477–503.
5. Došen, K. (2014). General proof theory. In Schroeder-Heister, P., Heinzmann, G., Hodges, W., and Bour, P. E., editors, *Logic, Methodology and Philosophy of Science: Proceedings of the 14th International Congress (Nancy 2011)*, pages 149–151. College Publications, London.
6. Došen, K. and Petrić, Z. (2001). The maximality of Cartesian categories. *Mathematical Logic Quarterly*, 47:137–144.
7. Dummett, M. (1975). The justification of deduction. *Proceedings of the British Academy*, pages 201–232. Separately published by the British Academy 1973. Reprinted in Dummett, M.: Truth and Other Enigmas, London: Duckworth 1978.
8. Ekman, J. (1994). *Normal Proofs in Set Theory*. Unversity of Göteborg, Ph.D. thesis.
9. Feferman, S. (1975). Review of Prawitz's 'Ideas and results in proof theory (1971)'. *Journal of Symbolic Logic*, 40:232–242.
10. Gentzen, G. (1934/35). Untersuchungen über das logische Schließen. *Mathematische Zeitschrift*, 39:176–210, 405–431 (English translation in: *The Collected Papers of Gerhard Gentzen* (ed. M. E. Szabo), Amsterdam: North Holland (1969), pp. 68–131).
11. Gentzen, G. (1969). *The Collected Papers of Gerhard Gentzen (ed. M.E. Szabo)*. North-Holland, Amsterdam.
12. Hilbert, D. (1918). Axiomatisches Denken. *Mathematische Annalen*, 78:405–415 (English translation in W. Ewald (ed.), From Kant to Hilbert. A Source Book in the Foundations of Mathematics, Vol. II, Clarendon Press, Oxford, 1996, pp. 1105–1115).
13. Hilbert, D. (1922). Neubegründung der Mathematik: Erste Mitteilung. *Abhandlungen aus dem Seminar der Hamburgischen Universität*, 1:157–177 (English translation in W. Ewald (ed.),

[9] Martin-Löf goes even further in establishing a relationship between logic and ethics [20], something which already German constructivism had argued for [17, 18].

From Kant to Hilbert. A Source Book in the Foundations of Mathematics, Vol. II, Clarendon Press, Oxford, 1996, pp. 1115–1134).
14. Hindley, J. R. (1997). *Basic Simple Type Theory*. Cambridge University Press, Cambridge.
15. Hintikka, J. (1975). *The Intentions of Intentionality and Other New Models for Modalities*. Reidel, Dordrecht.
16. Kreisel, G. (1971). A survey of proof theory II. In Fenstad, J. E., editor, *Proceedings of the Second Scandinavian Logic Symposium*, pages 109–170. North-Holland, Amsterdam.
17. Lorenzen, P. (1967). Moralische Argumentationen im Grundlagenstreit der Mathematiker. In *Tradition und Kritik. Festschrift für Rudolf Zocher*, pages 219–227. Stuttgart (repr. in P. Lorenzen, Methodisches Denken, Frankfurt: Suhrkamp 1968).
18. Lorenzen, P. and Schwemmer, O. (1975). *Konstruktive Logik, Ethik und Wissenschaftstheorie*. Bibliographisches Institut, 2nd ed., Mannheim.
19. Martin-Löf, P. (1984). *Intuitionistic Type Theory*. Bibliopolis, Napoli.
20. Martin-Löf, P. (2019). Logic and ethics. In Piecha, T. and Schroeder-Heister, P., editors, *Proof-Theoretic Semantics: Assessment and Future Perspectives. Proceedings of the 3rd Tübingen Conference on Proof-Theoretic Semantics, 27–30 March 2019*, pages 227–235. University of Tübingen (https://doi.org/10.15496/publikation-35319), Tübingen.
21. Moschovakis, Y. N. (1993). Sense and denotation as algorithm and value. In Oikkonen, J. and Väänänen, J., editors, *Logic Colloquium '90: ASL Summer Meeting in Helsinki*, pages 210–249. Springer, Berlin.
22. Piecha, T. and Schroeder-Heister, P., editors (2016). *Advances in Proof-Theoretic Semantics*. Springer, Cham.
23. Prawitz, D. (1965). *Natural Deduction: A Proof-Theoretical Study*. Almqvist & Wiksell, Stockholm. Reprinted Mineola NY: Dover Publ., 2006.
24. Prawitz, D. (1971). Ideas and results in proof theory. In Fenstad, J. E., editor, *Proceedings of the Second Scandinavian Logic Symposium (Oslo 1970)*, pages 235–308. North-Holland, Amsterdam.
25. Prawitz, D. (1972). The philosophical position of proof theory. In Olson, R. E. and Paul, A. M., editors, *Contemporary Philosophy in Scandinavia*, pages 123–134. John Hopkins Press, Baltimore, London.
26. Prawitz, D. (1973). Towards a foundation of a general proof theory. In Suppes et al., P., editor, *Logic, Methodology and Philosophy of Science IV*, pages 225–250. North-Holland.
27. Prawitz, D. (1974). On the idea of a general proof theory. *Synthese*, 27:63–77.
28. Quine, W. V. (1969). Speaking of objects. In *Ontological Relativity and Other Essays*, pages 1–25. Columbia University Press, New York.
29. Schroeder-Heister, P. (2006). Validity concepts in proof-theoretic semantics. *Synthese*, 148:525–571. Special issue *Proof-Theoretic Semantics*, edited by R. Kahle and P. Schroeder-Heister.
30. Schroeder-Heister, P. (2012/2018). Proof-theoretic semantics. In Zalta, E., editor, *Stanford Encyclopedia of Philosophy*. http://plato.stanford.edu, Stanford.
31. Schroeder-Heister, P. (2016). Open problems in proof-theoretic semantics. In Piecha, T. and Schroeder-Heister, P., editors, *Advances in Proof-Theoretic Semantics*, pages 253–283. Springer, Cham.
32. Schroeder-Heister, P. and Tranchini, L. (2017). Ekman's paradox. *Notre Dame Journal of Formal Logic*, 58:567–581.
33. Sommaruga, G. (2000). *History and Philosophy of Constructive Type Theory*. Kluwer, Dordrecht.
34. The Univalent Foundations Program (2013). *Homotopy Type Theory: Univalent Foundations of Mathematics*. Institute for Advanced Study, Princeton.
35. Thiele, R. (2003). Hilbert's twenty-fourth problem. *American Mathematical Monthly*, 110 (January):1–24.
36. Thiele, R. and Wos, L. (2002). Hilbert's twenty-fourth problem. *Journal of Automated Reasoning*, 29:67–89.

37. Tranchini, L. (2016). Proof-theoretic harmony: towards an intensional account. *Synthese*, (https://doi.org/10.1007/s11229-016-1200-3).
38. Tranchini, L., Pistone, P., and Petrolo, M. (2019). The naturality of natural deduction. *Studia Logica*, 107:195–231.
39. Wittgenstein, L. (1958). *Philosophical Investigations*. Basil Blackwell, Oxford. edited by Anscombe, G.E.M. and Rhees, R.

Peter Schroeder-Heister is Professor of Logic and Philosophy of Language as well as of Theoretical Computer Science at the University of Tübingen, where he taught from 1989 until 2019. One of his main areas of research is general proof theory and its philosophical basis, for which he has coined the term "proof-theoretic semantics". Further fields of interest include the history and philosophy of logic and mathematics, especially constructive and non-classical approaches to logical reasoning.

Chapter 9
Proofs as Objects

Wilfried Sieg

> The objects of proof theory shall be the proofs carried out in mathematics proper.

Abstract The rigor of mathematics lies in its systematic organization that supports *conclusive proofs* of assertions on the basis of assumed principles. *Proofs* are constructed through thinking, but they can also be taken as objects of mathematical thought. That was the insight prompting Hilbert's call for a "theory of the specifically mathematical proof" in 1917. This pivotal idea was rooted in revolutionary developments in mathematics and logic during the second half of the 19-th century; it also shaped the new field of *mathematical logic* and grounded, in particular, Hilbert's *proof theory*. The derivations in logical calculi were taken as "formal images" of proofs and thus, through the *formalization of mathematics*, as tools for developing a theory of mathematical proofs. These initial ideas for proof theory have been reawakened by a confluence of investigations in the tradition of Gentzen's work on natural reasoning, interactive verifications of theorems, and implementations of mechanisms that search for proofs. At this intersection of proof theory, interactive theorem proving, and automated proof search one finds a promising avenue for exploring the structure of mathematical thought. I will detail steps down this avenue: the formal representation of proofs in appropriate mathematical frames is akin to the representation of physical phenomena in mathematical theories; an important dynamic aspect is captured through the articulation of bi-directional and strategically guided procedures for constructing proofs.

The motto is a deeply programmatic remark in Gentzen (1936, 499), Gentzen's classical paper in which he proved the consistency of elementary arithmetic by transfinite induction up to ϵ_0. It fully coheres with remarks by Hilbert, as we will see.

W. Sieg (✉)
Department of Philosophy, Carnegie Mellon University, Pittsburgh, PA 15213, USA
e-mail: sieg@cmu.edu

F. Ferreira et al. (eds.), *Axiomatic Thinking I*,
https://doi.org/10.1007/978-3-030-77657-2_9

9.1 Introduction

In late December of 1933, Gödel delivered an invited lecture at the meeting of the Mathematical Association of America in Cambridge (Massachusetts); its title was, *The present situation in the foundations of mathematics.* Understanding by mathematics "the totality of the methods of proof actually used by mathematicians", Gödel articulated two problems for any foundation of mathematics. The first problem asks to state the methods of proof as precisely as possible and to reduce them to a minimum; the second seeks to justify the axioms involved and to provide "a theoretical foundation of the fact that they lead to results agreeing with each other and with empirical facts". Then Gödel asserted that the first problem "has been solved in a perfectly satisfactory way" and that the solution consists "in the so-called *formalization* of mathematics,"

> which means that a perfectly precise language has been invented, by which it is possible to express any mathematical proposition by a formula. Some of these formulas are taken as axioms, and then certain rules of inference are laid down which allow one to pass from the axioms to new formulas and thus deduce more and more propositions, the outstanding feature of the rules of inference being that they are purely formal, i.e., refer only to the outward structure of the formulas, not to their meaning, so that *they could be applied by someone who knew nothing about mathematics, or by a machine.* [My emphasis; cf. also Poincaré's remarks in Note 12.]

Today, these observations seem to be almost quaint and too obvious to be worth quoting. They do describe, however, the endpoint of a remarkable evolution away from an influential philosophical perspective that required proofs to be accompanied by the intuition of the mathematical objects they presumably deal with.[1]

How is it that the mathematical activity of proving theorems can be freed from that requirement and be represented as formal manipulation of symbolic configurations in accord with fixed rules? I will argue that the steps toward formalization are grounded in the radical transformation of mathematics in the second half of the 19-th century and the contemporaneous dramatic expansion of logic. In the Preface to the first edition of his essay *Was sind und was sollen die Zahlen?*, Dedekind asserts, "This essay can be understood by anyone who possesses, what is called common sense; ..." Dedekind emphasizes then immediately, and articulates in the spirit of the last sentence of Gödel's remark, that philosophical or mathematical "school knowledge" is not in the least needed for understanding it.[2]

What is needed, however, is the capacity to *use the characteristic conditions* of precisely defined concepts in the *stepwise construction* of arguments, i.e., of proofs. The steps in proofs are *not* to appeal to "inner intuition" and, according to Dedekind, that condition is imposed by the nature of our *Treppenverstand*. Proofs, understood

---

[1] That is a requirement of Kant's. It is discussed in detail and with reference to the logicist tendencies of Dedekind and Hilbert in Sieg (2016).

[2] Dedekind (1932, 336) In German, "Diese Schrift kann jeder verstehen, welcher das besitzt, was man den gesunden Menschenverstand nennt; philosophische oder mathematische Schulkenntnisse sind dazu nicht im geringsten erforderlich.".

in this way, play a central role for Dedekind as witnessed by his demand in the very first sentence of the Preface, "Was beweisbar ist, soll in der Wissenschaft nicht ohne Beweis geglaubt werden."[3] Two features of proofs are crucial for making plausible Dedekind's initial assertion concerning the understanding of his essay, namely, (1) the exclusive focus on characteristic conditions of concepts as starting-points and (2) the exclusive use of elementary steps for obtaining logical consequences. I point out that Hilbert called the characteristic conditions of fundamental concepts "axioms".

The first feature is examined in Sect. 9.2 under the heading *Mathematical context: structural concepts*. It will be illustrated by pertinent examples from Dedekind's foundational essays and Hilbert's early work. The second feature is described in Sect. 9.3, *Logical analysis: natural formal proofs*. Following Whitehead and Russell, Hilbert started in 1917 to systematically use and mathematically sharpen the formal logical tools of *Principia Mathematica*. However, the idea of using these tools to articulate difficult epistemological problems was pivotal. The problems included the decision problem, but also the call for a "theory of the specifically mathematical proof".[4] That call began to be answered only in the winter term 1921–22. That semester saw the invention of proof theory, the first steps in pursuit of the finitist consistency program, and the formulation of a new logical calculus that aimed for a more direct representation of elementary logical steps. About ten years later, this new *axiomatic calculus* of Hilbert and Bernays was transformed by Gentzen into a *rule-based natural deduction calculus* and that, in turn, is expanded into the *normal intercalation calculus*. The latter is used for the *bi-directional* construction of proofs.[5]

Section 9.4, *Natural formalization: CBT*, sketches the formal verification of the Cantor-Bernstein Theorem from the axioms of ZF.[6] It broadens the rule-based approach from logical connectives to mathematical definitions and insists that rule applications are always goal-directed and strategically guided. Thus, definitions are systematically incorporated in the bi-directional construction of proofs. Crucial is

[3] In English, "What is provable should not be believed in science without proof." An underlying principled separation of *analysis* (leading to fundamental concepts) and *synthesis* (using those concepts as the sole starting-points for the development of a subject) is articulated for elementary number theory most clearly in Dedekind's letter to Keferstein (Dedekind 1890). The significance of creating new concepts is dramatically pointed out in the Preface to (Dedekind 1888, 339). For the broader methodological context, Dedekind points there to his *Habilitationsrede* (Dedekind 1854).

[4] Hilbert viewed the decision problem, i.e., the problem of deciding a mathematical question in finitely many steps, as the "best-known and the most discussed" question. This issue, he asserts, "goes to the essence of mathematical thought". (Hilbert 1918, 1113).

[5] Bernays started to work as Hilbert's assistant in the fall of 1917. He was intimately involved in every aspect of Hilbert's foundational work. In his 1922-paper, the integration of structural and formal axiomatics is expressed very clearly. He views, fully aligned with Hilbert's perspective, the representation of mathematical proofs in formalisms as a tool for their investigation not as a way of characterizing mathematics as a formal game. About the logical calculus of "Peano, Frege, and Russell" he writes: these three logicians expanded the calculus in such a way "that the thought-inferences of mathematical proofs can be completely reproduced by symbolic operations." (p. 98).

[6] The full verification is presented in Sieg and Walsh (2019) and discussed with a particular focus on "proof identity" in Sieg (2019); see corresponding remarks in Sect. 9.5 below.

also a hierarchical conceptual organization of the material for obtaining a *mathematical frame* that facilitates the use of lemmas-as-rules. The totality of these tools reflect central aspects of ordinary mathematical practice in a completely natural way. Section 9.5, *Beyond formal verification*, examines the efficacy of these tools in the automated search for humanly intelligible proofs. As a case of going "beyond", I will discuss the search for proofs of Gödel's incompleteness theorems.

Sections 9.4 and 9.5 bring to life both features of Dedekind's perspective on proofs and enrich them in a modern way with computational experimentation. As far as the first feature is concerned, Saunders MacLane suggested in his (1934) that structural concepts are obtained by detailed *analyses of proofs*. As far as the second aspect is concerned, Dedekind himself hoped that his essay might stimulate other mathematicians to reduce the "long sequences of inferences to a more moderate, more pleasant size" (Dedekind 1888, 338). Clearly, changes of mathematical frames as well as of proof strategies can be implemented and set to work for experiments in a computational environment.

Ever since Gödel's second incompleteness theorem revealed the profound difficulty of the finitist consistency problem and Gentzen's calculi offered tools to address it, proof theorists have been preoccupied with the investigation of *formal deductions* in theories for arithmetic, subsystems of analysis or parts of set theory. In order to make their work relevant for mathematical practice, it has of course been important that mathematical proofs *can be formalized* in those theories, but questions like "*How* are they formalized?", "Do they have *structural features of practical significance*?", "Are these structural features of *broader methodological interest*?", and "Do they reveal *crucial aspects of mathematical cognition*?"—such questions have not been topics of detailed investigations.

In this paper, I am taking tentative steps toward finding ways to answer such questions. If mathematics is, as Gödel suggested in 1933, "the totality of the methods of proof actually used by mathematicians", then a third problem has to be raised—in addition to the two problems Gödel articulated for any foundation. An adequate representation of proofs cannot be achieved by using the minimum of formal methods. Rather, it has to be given in appropriate formal frames akin to the representation of physical phenomena in mathematical theories; important dynamic aspects are captured through the articulation of systematic procedures for constructing proofs.

9.2 Mathematical Context: Structural Concepts

Mathematical proofs are constructed through thinking and can be objects of mathematical thought. That was Hilbert's idea already in 1900 when he formulated the *24-th Problem* for his Paris talk; see Thiele (2003). Ultimately, he may have viewed the problem as too open-ended, since it is not included in his final list of *Mathematical Problems*. The problem requested:

> Develop a theory of the method of proof in mathematics in general.

The request is followed by a remark on the simplicity of proofs that claims, "under a given set of conditions there can be but one simplest proof". Hilbert continues,

> Quite generally, if there are two proofs for a theorem, you must keep going until you have derived each from the other, or until it becomes quite evident what variant conditions (and aids) have been used in the two proofs.

The development of an investigation of proofs began only after Hilbert had addressed the Swiss Mathematical Society in Zürich on 11 September 1917 and demanded in his talk *Axiomatisches Denken*,

> … we must - that is my conviction - turn the concept of the specifically mathematical proof into an object of investigation, just as the astronomer considers the movement of his position, the physicist studies the theory of his apparatus, and the philosopher criticizes reason itself.

Hilbert admitted immediately that "the execution of this program is at present, …, still an unsolved task".[7]

At the time of this talk, Hilbert was 55 years old and one of the world's most distinguished mathematicians. He had done groundbreaking work in core areas of mathematics during the 1890s that culminated in his *Zahlbericht*; he had pursued work on the foundations of geometry and analysis that was published in *Grundlagen der Geometrie* and in *Über den Zahlbegriff*; he had lectured since the mid-1890s on various parts of physics and in 1915 he had contributed to the theory of general relativity. This remarkable experience in mathematics and its "applications" is reflected in his Zürich talk.

Hilbert begins his talk by discussing the relations of mathematics to the two "great realms" of physics and epistemology. He claims that the essence of these relations becomes especially clear when describing the *general research method*, "which seems to have become more and more prominent in the newer mathematics", the *axiomatic method*. Why is this method becoming prominent in the "newer" mathematics? Isn't its *classical* paradigm already found in Euclid's *Elements*? Does Hilbert's *Grundlagen der Geometrie* not follow that very method and present the mathematical core of the *Elements* without the "flaws" that had been discovered in the 19-th century?

"Yes", Hilbert's *Festschrift* does take into account the corrections that had been proposed for a logically flawless foundation, but also most definitely "No": it is not an improved version of the *Elements* as an exposition of geometry. It is rather a penetrating reflection on the conceptual organization of the field and on metamathematical issues like independence and relative consistency. The axiomatic system is set up as a *structural definition* articulating the properties of the fundamental geometric notions that characterize relations between three unspecified systems of things:

> We think three different systems of things: we call the things of the first system *points* …; we call the things of the second system *lines* …; we call the things of the third system *planes*

[7] The call for such an investigation was not a whim for Hilbert. After all, we just saw that he had already formulated in 1900 the 24-th problem concerning "a theory of the method of proof in mathematics".

…; We think the points, lines, planes in certain mutual relations..; the precise and complete description of these relations is obtained by the *axioms of geometry*.

Hilbert's axioms do not express a priori truths but define the abstract notion of a *Euclidean Space* in the same way in which the axioms for groups or fields define the concept of a *group* or of a *field*.[8]

Hilbert followed in the footsteps of Dedekind who had introduced in *Was sind und was sollen die Zahlen?* the notion of a *simply infinite system* and considered as *natural numbers* the elements of any system falling under that notion. Here is Dedekind's formulation:

A system N is *simply infinite* if and only if there is an element 1 and a mapping ϕ, such that the characteristic conditions (α) – (δ) hold for them.

The characteristic conditions of the concept are: (α) ϕ[N] is contained in N, (β) N is the chain of the system {1} with respect to ϕ, (γ) 1 is not an element of ϕ[N], and (d) ϕ is injective.—Dedekind formulated *meta-mathematical problems* and achieved significant *results*: (1) He proved the "consistency" of the notion via the logically (and problematically) defined system ***N*** or, as he put it in his famous letter to Keferstein, he showed that the notion simply infinite system does not contain an "internal contradiction". (2) He proved a *representation theorem* showing that every simply infinite system is isomorphic to ***N*** and directly implying that the notion is *categorical*. (3) He argued for the *proof theoretic equivalence* of different simply infinite systems.[9]

The argument for the last result exploits the two features of Dedekind's approach I emphasized in the Introduction, namely, the requirement that in proofs *only* the characteristic conditions of the fundamental concepts are appealed to as starting points, and the conviction that the elementary logical steps are the same in every area of mathematics. This approach is prefigured in his earlier essay *Stetigkeit und irrationale Zahlen* for the concept of a complete ordered field and executed for real numbers in Hilbert's *Über den Zahlbegriff*. For both Dedekind and Hilbert, the nature of the objects in "models" does not enter proofs (Here, a "model" is just any system that falls under the structural concept; when arguing that a particular system falls under the structural concept, the nature of the objects plays of course a role.) In a different but complementary way, Hilbert made the same point in *Grundlagen der Geometrie* by giving an analytic model for geometry and a geometric model for analysis.

This perspective on mathematics is an absolutely radical conception of the subject. Dedekind and Hilbert locate their work in *logic*, broadly conceived. For Dedekind, mathematics *is* part of logic: the work in his two foundational essays, (1872) and (1888), gives a logical analysis of the number concept and develops the theory thereof systematically. For Hilbert, *Grundlagen der Geometrie* gives a "logical analysis of

[8] See my paper (Sieg 2014), in particular, the analysis of the Frege-Hilbert correspondence.

[9] The notion of proof theoretic equivalence was introduced in Sieg and Morris (2018) and is based on a close reading of #73 and #134 in Dedekind (1888), as well as the study of the penultimate version of that essay.

our spatial intuition" and investigates the question of what can be proved from what—without appealing to the intuition that is being analyzed. The issue to be addressed next is, what inferential principles can be used in proofs?

9.3 Logical Analysis: Natural Formal Proofs

On account of the fact that Dedekind had not given an explicit list of inferential steps, Frege severely and polemically criticized the former's essay (1888); that critique clearly extends to Hilbert's later *Grundlagen der Geometrie.* In the Preface of his *Grundgesetze der Arithmetik*, Frege claimed that the brevity of Dedekind's development of arithmetic in (1888) is only possible, "because much of it is not really proved at all". He continued:

> ... nowhere is there a statement of the logical or other laws on which he builds, and, even if there were, we could not possibly find out whether really no others were used – for to make that possible the proof must be not merely indicated but completely carried out.[10]

Apart from making the logical principles explicit there is an additional aspect that is hinted at in Frege's critique and detailed below.[11] Frege thinks of his own work as standing in the Euclidean tradition but going beyond it, because it lists not only the axioms in advance, but also the inferential principles. Both are articulated in a precise and expressive artificial language. Furthermore, the inferential principles have to be applied in a rule-bound manner.[12] Frege asserted that in his logical system "inference is conducted like a calculation" and observed:

> I do not mean this in a narrow sense, as if it were subject to an algorithm the same as ... ordinary addition or multiplication, but only in the sense that there is an algorithm at all, i.e., a totality of rules which governs the transition from one sentence or from two sentences to a new one in such a way that nothing happens except in conformity with these rules. (Frege 1984, 237).

[10] *Grundgesstze der Arithmetik*, p. 139 of *Translations from the philosophical writings of Gottlob Frege*, Peter Geach and Max Black (eds.), Oxford 1977.

[11] Poincaré's 1902-review of Hilbert's *Grundlagen der Geometrie* brings out this additional aspect in a quite vivid way, namely, through the idea of formalization as machine executability: "M. Hilbert has tried, so-to-speak, putting the axioms in such a form that they could be applied by someone who doesn't understand their meaning, because he has not ever seen either a point, or a line, or a plane. It must be possible, according to him, to reduce reasoning to purely mechanical rules." Indeed, Poincaré suggests giving the axioms to a reasoning machine, like Jevons' logical piano, and observing whether all of geometry would be obtained. Such formalization might seem "artificial and childish", were it not for the important question of completeness: "Is the list of axioms complete, or have some of them escaped us, namely those we use unconsciously? ... One has to find out whether geometry is a logical consequence of the explicitly stated axioms, or in other words, whether the axioms, when given to the reasoning machine, will make it possible to obtain the sequence of all theorems as output [of the machine].".

[12] It is a normative requirement (to insure intersubjectivity on a minimal cognitive basis), but it is also a practical one: if an inferential step required a proof that its premises imply its conclusion, we would circle into an infinite regress. Frege pursued also the particular philosophical goal of gaining "a basis for deciding the epistemological nature of the law that is proved." (Grundgesetze, 118).

That points to one crucial element of modern logic—developed by Frege himself and, among others, Peano, Whitehead and Russell—that is the basis for *formally* presenting parts of mathematics. The question of what might be understood by a formal presentation or a formal theory led in the 1930s to the introduction of rigorous mathematical notions to characterize "formality", "algorithm" or "computation". The most familiar mathematical notions are Gödel's *general recursiveness*, Church's *λ-definability* and Turing's *machine computability.*[13]

Hilbert had argued already in his (1905) for the joint development of logic and mathematics in pursuit of the meta-mathematical goal of a "direct" proof for consistency. He did not get very far, as his view of symbolic logic in the modern sense was extremely limited. He began studying *Principia Mathematica* in 1913. That provided the background for taking up such a joint development in 1917 with the *full logical toolbox* of Whitehead and Russell's work. The latter was radically reshaped and extended in his lectures during the winter term 1917–18. These lectures make for dramatic reading, as one can see in them the invention and presentation of a new subject that Hilbert called *mathematical logic*. One goal of this mathematical logic was the *full formalization* of mathematical practice. In a rough and ready way, the lectures achieved that goal for parts of number theory and analysis. Hilbert recognized, however, that a more adequate representation of proofs required moving from the awkward calculus of *Principia Mathematica* toward a more *direct* method of formalization.

Hilbert and Bernays took a first step in that direction via a novel *axiomatic calculus* they introduced in early 1922. It was used in all their subsequent proof theoretic work extending to 1939, when the second volume of *Grundlagen der Mathematik* was published. The reasons for introducing this calculus and abandoning that of *Principia Mathematica* are partly methodological and partly pragmatic. Methodologically, the organization of the calculus into groups of axioms for each individual logical connective was to parallel that of the axiomatic system in *Grundlagen der Geometrie*, where groups of axioms were formulated for the basic concepts of geometry. Here are the axioms for conjunction and disjunction:

$$\text{Conjuction I}: \quad A \to (B \to A\&B)$$
$$\text{Conjuction E}: \quad A\&B \to A \qquad A\&B \to B$$
$$\text{Disjunction I}: \quad A \to A \vee B \qquad B \to A \vee B$$
$$\text{Disjunction E}: \quad A \vee B \to ((A \to C) \to ((B \to C) \to C))$$

Pragmatically, the new calculus was to provide a more natural and direct way of formalizing proofs by bringing out the meaning of the logical connectives in logical inference steps. For example, having proved the conjunction A&B, the step

[13] The adequacy of these notions for capturing the informal concepts is still discussed today under the special headings, Church's Thesis, or Turing's Thesis or the Church-Turing Thesis. In my Sieg (2018), I gave a structural-axiomatic characterization of the *concept* of computation. That turns the "adequacy problem" into a standard problem any mathematical concept has to face when confronted with the phenomena it is purportedly capturing.

to A would be mediated in this way: the proof of A&B is expanded by the axiom A&B $\rightarrow$ A, then A is inferred via the rule of modus ponens. (They consider a linear presentation of proofs; if A&B has been obtained in a proof, then A&B can be repeated at the end of the given proof.)[14] To reach a *natural formalization* was not only Hilbert's goal, but a few years later also Gentzen's.

Gentzen is the proof theorists' proof theorist. His investigations and insights concern prima facie formal proofs. However, when he specifies in his 1936-paper the concept of a *deduction* he adds in parentheses *formal image of a proof*, i.e., deductions are viewed as formal images of mathematical proofs and are obtained by formalizing the latter. The process of formalization is explained as follows:

> The words of ordinary language are replaced by particular *signs*, the logical inference steps [are replaced by] rules that form new formally presented statements from already proved ones.

Only in this way, Gentzen claims, is it possible to obtain a "rigorous treatment of proofs". As if to make that point crystal clear, he presents in minute detail an informal proof of the infinity of primes using as inferential principles only steps that turn out to be rules of his natural deduction calculus.[15] Returning to the remark on deductions as formal images of mathematical proofs, he emphasizes, "*The objects of proof theory shall be the proofs carried out in mathematics proper*." (Gentzen 1936, 499).

The calculus of Gentzen's (1936) is a natural deduction one and goes back to his thesis. It is a version of Hilbert and Bernays' calculus in which the axioms have been transformed into I(ntroduction)- and E(limination)-rules for the logical connectives and to which one crucial new feature has been added: *making and discharging assumptions*. This 1936-paper used, in Gentzen's own terminology, a *sequent formulation of natural deduction*, i.e., at each node of a proof tree there is a sequent configuration $\Gamma \supset \phi$. In the top-down, forward construction of proofs all the directly logical actions happen on the r.h.s. of the sequent symbol $\supset$; the finite set Γ of formulas keeps only track of the assumptions on which the proof of ϕ depends. Notice that in this way of building proofs, one can make "detours" by inferring a formula by an I-rule and immediately applying the corresponding E-rule. Proofs without detours are called *normal*.[16]

Let me emphasize most strongly that Gentzen viewed making and discharging assumptions as *the* characteristic feature for his natural deduction calculi and as reflecting an *absolutely fundamental aspect of* proof construction in *mathematical*

---

[14] Proofs officially are sequences of formulas, but for the proof theoretic investigations, Hilbert and Bernays turn them into *proof trees* by a process they call "Auflösung in Beweisfäden".

[15] It is unfortunate that "Kalkül des natürlichen Schließens" has been translated as "natural deduction calculus". "Calculus of natural reasoning" would express better that it is a calculus reflecting in a formal way patterns of natural, informal argumentation.

[16] Gentzen's *Urdissertation* (Gentzen 1932–33) contains a formulation of a natural deduction calculus for intuitionist logic. Gentzen proved for that logic a normalization theorem; see (von Plato 2008) and (Sieg 2009). The systematic investigations of both classical and intuitionist natural deduction calculi was taken up in (Prawitz 1965); Prawizt established normalization theorems and discovered important structural features of normal proofs.

practice. Another general move in the construction of informal proofs is, however, not reflected in the syntactic configurations Gentzen considered. The additional general move is *taking backward steps*, for example, in proofs of universally quantified assertions or in indirect arguments. So, the question is, how can backward steps be joined with forward steps in a *single syntactic configuration*?

For the purpose of obtaining a single syntactic configuration, Gentzen's *sequent presentation* of natural deduction is a convenient starting-point. It can be modified to an *IC calculus*[17] that allows bi-directional reasoning through *intercalating* formulas and *using structured sequents* of the form $\Gamma;\ \Delta \supset \phi$. Γ still has the role of containing the assumptions on which ϕ depends, whereas Δ consists of formulas that are obtained by E-rules, (successively) applied to elements of Γ. The formula ϕ on the r.h.s. of the sequent symbol $\supset$ is called *goal*. Forward steps toward the goal are taken via E-rules on the l.h.s., whereas backward steps from the goal are reflected by applications of inverted I-rules on the r.h.s.. Notice that the E-rules yield as consequences only (strictly) positive subformulas of their premises. In sum, the forward steps make it possible to *extract* more specific information from assumptions and backward steps lead by *inversion* to less complex new proof obligations.

A modification of the IC calculus, its normal *NIC* version, restricts forward steps to be *goal-directed* in the following sense: elimination rules are applied only if the goal formula ϕ is a (strictly) positive subformula of an assumption in Γ. Here is a very simple proof that involves only conjunction but allows the reader, nevertheless, to discover the fundamental ideas.

$$\dfrac{\dfrac{\dfrac{(A\&B)\&C;\ A\&B,\ A \supset A}{(A\&B)\&C;\ A\&B \supset A}}{(A\&B)\&C \supset A} \qquad \dfrac{(A\&B)\&C;\ C \supset C}{(A\&B)\&C \supset C}}{(A\&B)\&C \supset A\&C}$$

Remark The evolution of natural deduction calculi and their diagrammatic presentation is described in great detail in Sect. 1 of Sieg and Derakhshan (2021). That description includes, in addition, an important discussion of the diagrams that were used by Jaskowski and Fitch. Indeed, the representation of bi-directionally constructed natural deduction proofs is given by *restricted Fitch-diagrams*. The above proof construction would begin with the configuration.

(A&B)&C	Premise
.	
A&C	Goal

[17] "Intercalate" does not only mean "interpolate an intercalary period in a calendar" but also "insert something between layers in a crystal lattice or other structure". So, I apply it to the insertion of formulas between layers in logical proof structures. ("Interpolate" could not be used in this logical context for obvious reasons.).

The task is to bridge the gap between Premise and Goal. The elimination strategy would not be used, as the Goal is not a positive subformula of the Premise. So,the inverted &-I would yield the next configuration:

(A&B)&C	Premise
.	
A	New Goal
.	
C	New Goal
A&C	Goal

The reader can undoubtedly take the next steps in this bi-directional proof construction to obtain this completed proof:

1. (A&B)&C	Premise
2. A&B	&El: 1
3. A	&El: 2
4. C	&Er: 1
5. A&C	&I: 3,4

This is a representation that is obviously much more economical and genuinely suitable for a computer screen, in particular, once subderivations are being considered in a quasi-linear presentation of "boxes" within "boxes" as it is done via Fitch diagrams. For human construction on paper it is most inconvenient, whereas the computer's capacities can be effectively exploited for this naturally bi-directional construction. One can find in Sieg and Walsh (2019) complex examples. Perhaps it should be emphasized that the finished proof can be presented with annotations that reflect the order of the steps taken for obtaining this proof.

NIC proofs are easily translated into normal natural deduction proofs. Their *construction* is strategically guided to close the gap between assumptions and goals by intercalating steps. The strategies reflect, on the one hand, guidance for informal proof construction and, on the other hand, rely on structural features of normal proofs.[18] They are very efficient for the construction of proofs in logic, but how can their use be expanded to formally represent proof construction in (parts of) mathematics?

---

[18] For the NIC calculus one can prove a *strengthened completeness theorem* for both classical and intuitionist logic: for any ϕ and Γ, there is either a *normal* proof of ϕ from Γ or a counterexample to Γ, ϕ. The central considerations in the completeness proof have been organized into efficient logical strategies for automated search and have been implemented in the system AProS. The completeness proofs are found in Sieg and Byrnes (1998) and Sieg and Cittadini (2005).

9.4 Natural Formalization: CBT

For the expansion of the basic approach of bi-directional argumentation to mathematics, the logical toolbox has to be enriched by new components:

(1) the specification of *definitions* by E- and I-rules extending the rule-based approach from logical connectives to mathematical operations and notions;
(2) a well-structured *conceptual organization* its *formal frame*;
(3) the integration of proofs with the formal frame by using *lemmas-as-rules*.[19]

These are the characteristic components of *natural* formalization; "natural" is used here in analogy to how it is used in *natural* deduction. They are used informally and formally.

Such an expansion was implemented in the system AProS. In Sieg and Walsh (2019), AProS was used as a proof checker to formally verify in ZF a classical result of set theory, the Cantor-Bernstein Theorem CBT:

$$\text{If } f \in \text{inj}(a, b) \text{ and } g \in \text{inj}(b, a), \text{ then a} \approx \text{b}.$$

Here, $f \in \text{inj}(a, b)$ expresses that f is an injection from a to b and $g \in \text{inj}(b, a)$ expresses similarly that g is an injection from b to a; the consequent of the conditional states that a and b are equinumerous, i.e., there is a bijection between a and b.

CBT is actually equivalent to the assertion I have come to call *Dedekind's Fundamental Lemma*:

$$\text{If } e \subseteq d \subseteq a \text{ and } a \approx e, \text{ then } a \approx d.$$

The proof of the Fundamental Lemma using CBT is direct. To obtain CBT from this Fundamental Lemma, we have to specify subsets of a that satisfy the premises of the Lemma (using the premises of CBT). That is done for lines 3, 4 and 5 by appealing to very straightforward lemmas that are indicated in the proof of CBT in the AProS interface:

---

[19] Lemmas have been used as rules in classical texts but, of course, also for the standard structuring of mathematical expositions. Classical examples are found in Book I of Euclid's *Elements* (for example, in the proof of the Pythagorean Theorem the earlier, very important theorem I.4 is being used in just that way), but also in the development of the theory of systems and mappings in Dedekind (1888).

1. $f \in$ inj(a, b)	*Premise*
2. $g \in$ inj(b, a)	*Premise*
3. 1($g \bullet f$) $\in$ bij(a, $g \bullet f[a]$)	*Theorem (Core12)*: 1, 2
4. $g[b] \subseteq a$	*Theorem (Func17)*: 2
5. $g \bullet f[a] \subseteq g[b]$	*Theorem (Comp11)*: 1, 2
6. $a \approx g[b]$	*Theorem (Fundamental Lemma)*: 3, 4, 5
7. $b \approx g[b]$	*Theorem (Equi4)*: 2
8. $a \approx b$	*Theorem (Equi8)*: 6, 7

Proof of CBT in the AProS interface.

The proof crucially appeals to Dedekind's Fundamental Lemma as a rule and also to a few straightforward and individually meaningful lemmas of elementary set theory, e.g., line 8 is obtained from lines 6 and 7 by appealing to the transitivity of equinumerosity. This proof is based on a theoretical analysis of mathematical proofs and gives, as MacLane would suggest (1934, 57), "an adequate logical explanation of the fact, that genuine mathematical proofs can be given through short descriptions".

Dedekind proved the Fundamental Lemma and, from it, CBT in 1887; he never published the proofs. Zermelo proved CBT in 1908 without knowing of Dedekind's proof, but being quite familiar with Dedekind's notion of the *chain* of a system a w.r.t. a mapping f.[20] Such chains k satisfy a central *structural identity*, namely, $k = a \cup f[k]$. Dedekind and Zermelo considered in their respective proofs two different chains c and c^*. Instantiated with c and c^*, the structural identities are used to define two partitions of a and d, as well as bijections between their respective components; the unions of these bijections are then bijections between a and d.

The overall organization of the proofs is depicted in the diagram below. In Sieg and Walsh (2019), everything is proved in ZF from the ground up and is literally identical except for the parts of the proofs represented by the diamond in the diagram. I consider it as a paradigmatic case of natural formalization that includes, of course, the detailed analysis of mathematical proofs as a crucial component.

[20] It makes explicit the informal notion of *objects obtained from elements in a by finitely iterating f*; it is defined as the intersection of all sets that contain a as a subset and are closed under f.

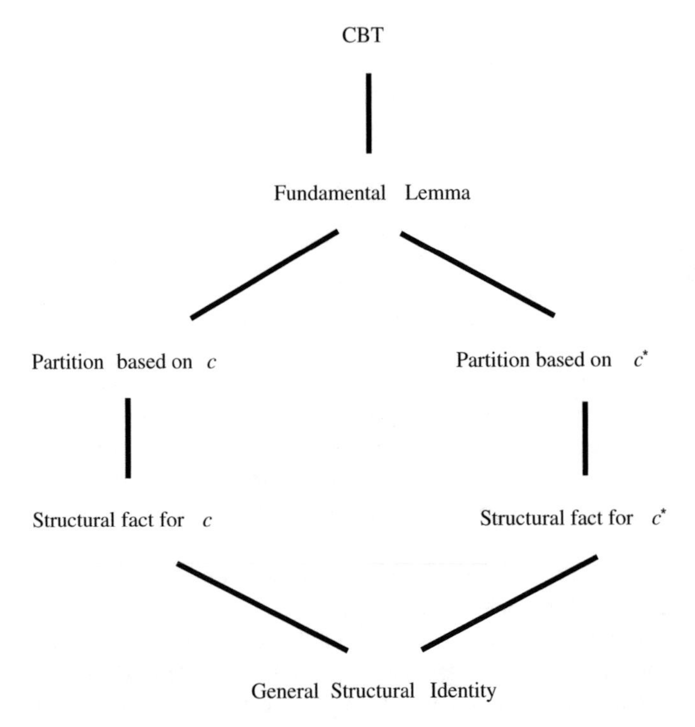

This is only one direction for investigating mathematical proofs. Other directions can be pursued. Recall Hilbert's request when he articulated his 24-th problem:

> Quite generally, if there are two proofs for a theorem, you must keep going until you have derived each from the other, or until it becomes quite evident what variant conditions (and aids) have been used in the two proofs.

Concerning Dedekind's and Zermelo's proofs the following question can be raised, *Are they at all different* or *are they identical*? In 1932, when these proofs could be compared for the first time, Emmy Noether viewed them as "exactly the same", whereas Zermelo thought of them as "inessentially different". In Hilbert's spirit, we can gain a mathematical insight from the comparison and see very clearly "what variant conditions (and aids) have been used in the two proofs": in Dedekind's proof, the chain c is the smallest fixed-point of the general structural identity $k = a \cup f\{k\}$, whereas in Zermelo's proof $c*$ is the largest.

Proof identity has been and will continue to be a central topic for any theory of mathematical proofs. In the case of CBT, I am convinced that all its known proofs are either of the Dedekind or Zermelo variety.[21] So, we have gained material that can inform a theory of mathematical proofs. For example, it seems clear that proof identity—tentatively taken as literal syntactic identity of their formalizations—must be

[21] That point is at the center of my paper (Sieg 2019). Aries Hinkis' book *Proofs of the Cantor-Bernstein Theorem: A mathematical excursion* is a comprehensive discussion of proofs and their history. I have examined most of the proofs in that mathematical excursion, with special care given to the proofs of König and Banach. They all fall into one of these categories.

a notion that depends both on the mathematical frame and on the strategies for proof construction. In the next section, I will introduce another source of complementary material, the automated search for *humanly intelligible formal proofs*.

9.5 Beyond Formal Verification

The focus on human-centered automated proof search has a long tradition in computer science. I am thinking, in particular, of Woody Bledsoe's work in the 1970s and 1980s. This tradition seems, at the moment, to be dormant in computer science. However, the mathematician Gowers has expressed a strong interest in what he calls "extreme human-centered automatic theorem proving"; see the *Interview with Sir Timothy Gowers*, (Diaz-Lopez 2016). A strategically guided automatic theorem prover, let's call it the *G&G-prover*, is presented in his joint paper with Ganesalingam (2017) and its earlier version (Ganesalingam and Gowers 2013). Their *central goal* is to generate, with the help of the automatic prover, a proof in English; the generated proof text is programmatically to be indistinguishable from excellent mathematical writing. That's feasible only, they rightly argue, if the automatic prover uses strategies that are also used by human mathematicians as every formal step taken by the prover is *paraphrased directly* in English. The latter is a decision on their *basic method* that is supposed to reflect the simple "rhetorical structure" of mathematical arguments, constituted by linearly organized forward steps.

Their central goal, basic method and observation on human-like strategies limit the moves their prover can make. Indeed, Ganesalingam and Gowers state that their prover, on account of the constraints on moves, is solving only "routine problems". Here are a few of the restrictions imposed: no backtracking moves are allowed, as they would interfere with the linear step-by-forward-step writing of proofs; the first-order language they are using does not contain negation; the treatment of disjunction is described as "work in progress" and will definitely require backtracking. In Sieg and Derakhshan (2021) the G&G-prover is analyzed and compared with AProS. It is shown that the G&G-prover's mathematical work can easily be mimicked in a very weak fragment of AProS. Nevertheless, I want to emphasize that there is a common perspective, namely, that mechanisms for human-centered automated proof search should "think" like mathematicians.[22]

I will turn now to a second example of a detailed analysis and significant reshaping of a complex proof that, ultimately, provided the tools for a successful automated search. During the last two years of my graduate studies at Stanford, I was working as a research associate at Patrick Suppes' IMSSS (Institute of Mathematical Studies in

[22] Gowers' and my work actually have a common motivating ambition: a pedagogical one. Gowers' is articulated in Diaz-Lopez (2016); mine has propelled the development of a fully web-based course *Logic & Proofs*; see Sieg (2007). The AProS website has this URL http://www.phil.cmu.edu/projects/apros/. AProS forms the basis of a dynamic, interactive *proof tutor* in this introduction to logic. The course has as its primary goal teaching students strategic proof construction. *Logic & Proofs* has been completed by more than 12,000 students for credit at their home institution.

the Social Sciences). The Institute was then a hotbed of *computer assisted instruction*, CAI. The Suppes team had developed a computer-based course in ZF set theory. My task was to formalize the proofs of Gödel's incompleteness theorems and closely related ones, all for ZF. How could one accommodate on a small screen, I asked myself at the very beginning, the sheer notational complexity of dealing with the formal system ZF, its meta-mathematical description, the arithmetization of that description and, finally, its representation within ZF?—My answer was, in the end: "Omit arithmetization and, instead, *represent directly* the central syntactic notions!" The latter notions (including Substitution, Formula and Proof in ZF) are, after all, precisely given by elementary inductive definitions and structural recursions. With this strategic simplification I succeeded in formally verifying the incompleteness theorems and related ones, for example, Löb's Theorem.[23]

Many years later, around 2003, AProS was endowed with the two levels of argumentation used for my Gödel proofs, in the object-, respectively meta-theory. I formulated suitable principles connecting these two levels, ProvI and ProvE. The first principle of *provability introduction* allows the step from an object theoretic ZF-proof of ϕ to the meta-theoretic assertion that ϕ is provable in ZF; the second principle of *provability elimination* allows the complementary step from the meta-theoretic assertion that ϕ is provable in ZF to using ϕ as a theorem in an object theoretic proof. With these principles, the semi-representability of the theorem predicate and the defining bi-conditional of the Gödel sentence, AProS very quickly finds a proof of the first theorem that is absolutely canonical. Indeed, it does so also for the second incompleteness theorem and related theorems, in particular, Löb's Theorem; see Sieg and Field (2005).

I consider detailed, unifying proof analyses and natural formalizations as necessary steps toward an investigation of the notion of *mathematical proof*. We have to exploit the rich body of mathematical knowledge that *is systematic*, but that is also *structured for human intelligibility and discovery*. In this way, we can isolate creative elements in proofs and formulate suitable heuristics. Such work gets us closer to uncovering techniques of our thinking and to realizing Hilbert's "fundamental idea" for his proof theory, namely, "to describe the activity of our understanding, to make a protocol of the rules according to which our thinking actually proceeds." (Hilbert 1927).

The above work on natural formalization and automated proof search has given us insights: the particular representation of proofs as restricted Fitch-diagrams allows the systematic bi-directional construction of *normal* proofs and reflects mathematical practice; for more advanced parts of mathematics, an embedding in a conceptually organized mathematical frame is necessary. Going beyond the analysis and formalization of proofs, we can implement search procedures and run computer experiments, observing how strategically motivated modifications affect search. This way of proceeding is also a way of building cognitive models of proof construction that

[23] See Sieg (1978) and Sieg et al. (1981). Ironically, it is quite clear that Gödel himself used such a direct representation in his original argument for a version of the first incompleteness theorem; see Gödel's description of his discovery in Wang (1981, 654).

are deeply informed by the practice of mathematics and may reveal sophisticated capacities as well as real limitations of the human mathematical mind. In any event, at the intersection of proof theory, interactive theorem proving and automated proof search, we find ways for exploring the structure of the mathematical mind.

Let me end by listening to another voice that connects the rich past with this programmatic future: Saunders MacLane was one of the last logic students in Göttingen and a good friend of Gentzen's. He completed his thesis *Abgekürzte Beweise im Logikkalkul* in 1934. A year later, he published a summary in which he pointed out that proofs are not "mere collections of atomic processes, but are rather complex combinations with a highly rational structure". He reflected in 1979 on his early work in logic and remarked, "There remains the real question of the actual structure of mathematical proofs and their strategy. It is a topic long given up by mathematical logicians, but one which still—properly handled—might give us some real insight." Obviously, I share MacLane's hope. I also think that the work described above already has given us some insights; whether they are to be called "real" or not, I leave to the reader's judgment.[24]

References

1. Bernays, P. 1922 Hilberts Bedeutung für die Philosophie der Mathematik; *Die Naturwissenschaften* 4, 93–99.
2. Bledsoe, W. 1977 Non-resolution theorem proving; *Artificial Intelligence* 9, 1–35.
3. Bledsoe, W. 1983 The UT natural-deduction prover; Technical Report, Department of Computer Science, University of Texas, Austin.
4. Dedekind, R. 1854 Über die Einführung neuer Funktionen in der Mathematik; Habilitationsvortrag; In (Dedekind 1932, 428–438) and translated in (Ewald 1996, 754–762).
5. Dedekind, R. 1872 *Stetigkeit und irrational Zahlen*; Vieweg. Translated in (Ewald 1996, 765–779).
6. Dedekind, R. 1888 *Was sind und was sollen die Zahlen?*; Vieweg. Translated in (Ewald, 1996, 787–833).
7. Dedekind, R. 1890 Letter to H. Keferstein, Cod. Ms. Dedekind III, I, IV. Translated in van Heijenoort (editor), *From Frege to Gödel*, Harvard University Press 1967, 98–103.
8. Dedekind, R. 1932 *Gesammelte Mathematische Werke*, Vol. 3; Vieweg.
9. Diaz-Lopez, A. 2016 Interview with Sir Timothy Gowers, Notices of the American Mathematical Society 63 (9), 1026–1028.
10. Ewald, W. B. (Editor) 1996 *From Kant to Hilbert: Readings in the Foundations of Mathematics*; Oxford University Press.
11. Fitch, F. 1952 *Symbolic Logic: An Introduction*; Ronald Press, N.Y.
12. Frege, G. 1893 *Grundgesetze der Arithmetik, begriffsschriftlich abgeleitet*; Pohle Verlag.
13. Frege, G. 1980 *Gottlob Freges Briefwechsel*; edited by G. Gabriel, F. Kambartel, and C. Thiel, Meiner.
14. Frege, G. 1984 *Collected Papers on Mathematics, Logic, and Philosophy*; edited by B. McGuinness. Oxford University Press.

[24] Implicitly, I argued against an artificial opposition of informal and formal proofs. By incorporating the strategic, dynamic aspect of interactive proof construction into a fully automated search procedure, one obtains the means for exploring structural features of proofs and their construction.

15. Ganesalingam, M. and Gowers, T. 2013 A fully automatic problem solver with human-style output. arXiv:1309.4501.
16. Ganesalingam, M. and Gowers, T. 2017 A fully automatic theorem prover with human-style output; *J. Automated Reasoning* 58, 253–291.
17. Gentzen, G. 1932/3 "Urdissertation", Wissenschaftshistorische Sammlung, Eidgenössische Technische Hochschule, Zürich, Bernays Nachlass, Ms. ULS.
18. Gentzen, G. 1934/5 Untersuchungen über das logische Schließen I, II, *Mathematische Zeitschrift* 39, 176–210, 405–431
19. Gentzen, G. 1936 Die Widerspruchsfreiheit der reinen Zahlentheorie. *Mathematische Annalen* 112 (1), 493–565.
20. Gödel, K. 1933 The present situation in the foundations of mathematics; in S. Feferman e.a. (editors), *Gödel's Collected Works* vol. III, Oxford University Press, 1995, 36–53.
21. Hilbert, D. 1899 *Grundlagen der Geometrie*; Teubner.
22. Hilbert, D. 1900a Über den Zahlbegriff; *Jahresbericht der DMV* 8, 180–183.
23. Hilbert, D. 1900b Mathematische Probleme; *Nachrichten der Königlichen Gesellschaft der Wissenschaften*, 253–297.
24. Hilbert, D. 1905 Über die Grundlagen der Logik und der Arithmetik; in *Verhandlungen des Dritten Internationalen Mathematiker Kongresses*, Teubner, 174–185.
25. Hilbert, D. 1918 Axiomatisches Denken; *Mathematische Annalen* 78, 405–415.
26. Hilbert, D. 1927 Die Grundlagen der Mathematik; *Abhandlungen aus dem mathematischen Seminar der Hamburgischen Universität* (6), 65–85.
27. Hilbert, D. and Bernays, P. 1917–18 *Prinzipien der Mathematik*; in W. B. Ewald and W. Sieg (editors) *David Hilbert's Lectures on the Foundations of Arithmetic and Logic, 1917–1933*, Springer, 2013, 64–214.
28. Hilbert, D. and Bernays, P. 1921–22 *Grundlagen der Mathematik*; in W. B. Ewald and W. Sieg (editors) *David Hilbert's Lectures on the Foundations of Arithmetic and Logic, 1917–1933* Springer, 2013, 431–518.
29. Hilbert, D. and Bernays, P. 1939 *Grundlagen der Mathematik*, volume II; Springer.
30. Jaskowski, S. 1934 On the rules of suppositions in formal logic; *Studia Logica* 1, 4–32.
31. MacLane, S. 1934 *Abgekürzte Beweise im Logikkalkul*; Dissertation, Göttingen.
32. MacLane, S. 1935 A logical analysis of mathematical structure; *The Monist* 45, 118–130.
33. MacLane, S. 1979 A late return to a thesis in logic. In I. Kaplansky (editor), *Saunders MacLane — Selected Papers*, Springer, 63–66.
34. Poincaré, H. 1902 Review of (Hilbert 1899), *Bulletin des sciences mathématiques* 26, 249–272.
35. Prawitz, D. 1965 *Natural deduction – A proof-theoretical study*; Almqvist & Wiksell.
36. Sieg, W. 1978 *Elementary proof theory*; Technical Report 297, Institute for Mathematical Studies in the Social Sciences, Stanford 1978, 104 pp.
37. Sieg, W. 1992 *Mechanisms and Search – Aspects of Proof Theory*, Vol. 14. Associazione Italiana di Logica e sue Applicazioni.
38. Sieg, W. 2007 The AProS Project: Strategic thinking & computational logic; *Logic Journal of the IGPL* 15 (4), 359–368.
39. Sieg, W. 2009 Review of (von Plato 2008), *Mathematical Reviews* 2413304.
40. Sieg, W. 2010 Searching for proofs (and uncovering capacities of the mathematical mind); In S. Feferman and W. Sieg (editors), *Proof, Categories and Computations*, College Publications:189–215.
41. Sieg, W. 2014 The ways of Hilbert's axiomatics: structural and formal; *Perspectives on Science* 22(1), 133–157.
42. Sieg, W. 2016 On Tait on Kant and Finitism. *Journal of Philosophy* 112(5– 6), 274– 285.
43. Sieg, W. 2018 What is the *concept* of computation? In F. Manea, R.G. Millere, and D. Nowotka (editors), *Sailing routes in the world of computation*, CiE 2018, Lecture Notes in Theoretical Computer Science and General Issues 10936, Springer, 386–396.
44. Sieg, W. 2019 The Cantor–Bernstein theorem: How many proofs?; *Philosophical Transactions of the Royal Society A*, 377 (2140), 20180031.

45. Sieg, W. and Byrnes, J. 1998 Normal natural deduction proofs (in classical logic); *Studia Logica* 60, 67–106.
46. Sieg, W. and Cittadini, S. 2005 Normal natural deduction proofs (in non-classical logics); in D. Hutter and W. Stephan (editors) *Mechanizing mathematical reasoning* Lecture Notes in Computer Science 2605, Springer, 169–191.
47. Sieg, W. and Derakhshan, F. 2021 Human-centered automated proof search; Journal of Automated Reasoning 65:1153–1190.
48. Sieg, W. and Field, C. 2005 Automated search for Gödel's proofs, *Annals of Pure and Applied Logic* 133, 319–338.
49. Sieg, W., Lindstrom, I. and Lindstrom, S. 1981 Gödel's incompleteness theorems - a computer-based course in elementary proof theory; in P. Suppes (editor) *University-Level Computer-Assisted Instruction at Stanford 1968–80*, Stanford, 183–193
50. Sieg, W. and Morris, R. 2018 Dedekind's structuralism: creating concepts and deriving theorems; in E. Reck (editor), *Logic, Philosophy of Mathematics, and their History: Essays in Honor of W.W. Tait*, College Publication, 251–301.
51. Sieg, W. and Walsh, P. 2019 Natural formalization: deriving the Cantor-Bernstein Theorem in ZF; *Review of Symbolic Logic*, 1–35; doi:https://doi.org/10.1017/S175502031900056X.
52. Thiele, R. 2003 Hilbert's twenty-fourth problem; *American Mathematical Monthly* 110, 1–24.
53. Von Plato, J. 2008 Gentzen's proof of normalization for natural deduction; *Bulletin of Symbolic Logic* 14, 240–257.
54. Wang, H. 1981 Some facts about Kurt Gödel; *Journal of Symbolic Logic* 46, 653–659.
55. Zermelo, E. 1908 Untersuchungen über die Grundlagen der Mengenlehre I; *Mathematische Annalen* 65 (2), 261–281.

Wilfried Sieg is Patrick Suppes Professor of Logic and Philosophy at Carnegie Mellon University. He received his Ph.D. from Stanford University in 1977. From 1977 to 1985, he was Assistant and Associate Professor at Columbia University. In 1985, he joined the Carnegie Mellon faculty as a founding member of the University's Philosophy Department and served as its Head from 1994 to 2005. He is internationally known for mathematical work in proof theory, historical work on modern logic and mathematics, and philosophical essays on the nature of mathematics. Over the last three decades he has developed and implemented novel methods for the human-centered automated search for proofs in logic, but also elementary set theory. Sieg is a Fellow of the American Academy of Arts and Sciences and also a Fellow of the American Association for the Advancement of Science.

Chapter 10
Where Do Axioms Come From?

Craig Smoryński

Abstract The traditional view going back to the Greeks is that axioms are true and that truth is inherited from them through logical reasoning. The modern view, forcefully espoused by David Hilbert in a letter to Gottlob Frege of 29 December 1899 runs counter to this.

> I was very much interested in your sentence: 'From the truth of the axioms it follows that they do not contradict one another', because for as long as I have been thinking, writing, lecturing about these things, I have been saying the exact reverse: If the arbitrarily given axioms do not contradict one another, then they are true, and the things defined by the axioms exist. This for me is the criterion of truth and existence.[1]

The Greek view, held by Frege, is that one chooses evident truths about some subject as axioms and generates additional truths through logical reasoning. Hilbert more-or-less states that one chooses axioms arbitrarily and, should they be consistent, they are true about something (and so too are the theorems derived from them). Hilbert's statement seems shockingly extreme. I do remember in my youth reading about the "absolute freedom" mathematicians have in choosing axiom systems to study. And, indeed, modern algebra affords an immodest collection of such: semi-groups, cancellative semi-groups, abelian semi-groups, cancellative abelian semi-groups, groups, abelian groups, rings of various types, etc., etc., etc. But not all axiom systems are equal and some theories are deemed more significant and worthy of our attention than others. Thus, one wishes philosophically to reject Hilbert's use of the phrase "arbitrarily given". One can still hold the view that the significant axiomatisations arise by taking true assertions about something. I wish here to emphasise that this does not describe practice. Axioms are chosen for a variety of reasons; I

---

[1]Gottlob Frege, *Philosophical and Mathematical Correspondence*, University of Chicago Press, Chicago, 1980, p. 42.

C. Smoryński (✉)
429 S. Warwick, Westmont, IL 60559, USA

F. Ferreira et al. (eds.), *Axiomatic Thinking I*,
https://doi.org/10.1007/978-3-030-77657-2_10

can name seven of these: Truth, Necessity, Proof-Generation, Convenience, System Refinement, Analogy, and Pure Formalism. Let us discuss these in turn.

TRUTH. For two millennia, if you had asked any western philosopher for an example of an axiomatic theory with true axioms, the obvious answer would have been Euclid's geometry as presented in *The Elements*. Only one axiom of the theory, namely the parallel postulate, did not please everybody, but not because it was not true. Everyone believed it was true, just not as intuitively true as the remaining axioms. Thus there were many attempts to prove the postulate from the other axioms, or, at least, to find an intuitive replacement. In the 18th and 19th centuries attempts to prove the parallel postulate by showing how its rejection led to a contradiction resulted in apparently consistent, noneuclidean geometries and to the entertaining by no less a mathematician than Carl Friedrich Gauss of the notion that space might actually be noneuclidean, i.e., Euclid's geometry might not in fact be a true theory of space. For all practical purposes, however, Euclidean geometry is good enough and, as it fits our intuition better than noneuclidean geometry, we would like to keep the theory with its ostensibly false parallel postulate. What justification can we offer for this postulate?

NECESSITY. Euclid delayed using the parallel postulate until it was absolutely necessary. This does not, of course, mean that he had his doubts about the truth of the axiom, but it does set the precedent for accepting doubtful axioms and noting when they are needed, a practice most familiarly applied to the axiom of choice in its accompanying distinction between effective constructions (those not depending on choice) and non-effective constructions (those appealing to choice) in set theory.[2] Another famous example of an axiom justified by its necessity is Bertrand Russell's axiom of reducibility, an axiom he acknowledged was given not because it was true, but because it was necessary:

> That the axiom of reducibility is self-evident is a proposition which can hardly be maintained. But in fact self-evidence is never more than a part of the reason for accepting an axiom, and is never indispensable. The reason for accepting an axiom, as for accepting any other proposition is always largely inductive, namely that many propositions which are nearly indubitable can be deduced from it, and that no equally plausible way is known by which these propositions could be true if the axiom were false, and nothing which is probably false can be deduced from it...Infallibility is never attainable, and therefore some element of doubt should always attach to every axiom and to all of its consequences.[3]

The argument is, perhaps, specious and unphilosophical: "I need this axiom, everybody does it, and nothing is perfect". Nonetheless, necessity has been a reason for accepting certain axioms and an entire subfield of set theory, the study of large cardinals, is predicated on it. It was justified by Kurt Gödel when he pointed out that, as a consequence of his Incompleteness Theorems, higher cardinals proved new results in arithmetic and, indeed, shortened considerably proofs of known theorems. Thus,

---

[2] With respect to the axiom of choice, I am fond of pointing out that the French semi-intuitionists did not believe in it, yet (unknowingly) made necessary use of it in some of their proofs.

[3] Bertrand Russell and Alfred North Whitehead, *Principia Mathematica to *56*, Cambridge University Press, Cambridge, 1962, p. 59.

he concluded, large cardinals should be studied in the hope of proving new theorems and finding more accessible proofs of known theorems.

PROOF-GENERATION. We are all familiar with proof-generated concepts whereby one goes through a list of properties of a particular structure used in proving it to satisfy some theorem, and then generalises the result to all structures with the given properties. Thus, some results proven to hold of the integers hold for arbitrary rings and some for integral domains, etc. As explained in some detail by Gregory H. Moore,[4] set theory was initially axiomatised in this manner when in 1908 Ernst Zermelo, in response to criticism of his first proof of the Well-Ordering Theorem, isolated those properties of sets used in his new proof of the theorem. This list of properties is quite modest:

I. (Restricted Comprehension). For any set x and any "definite property" $P(y)$, the set $\{y \in x \mid P(y)\}$ exists.
II. (Power Set). The set of all subsets of a given set, $P(x) = \{y \mid y \subseteq x\}$, exists.
III. (Intersection). For any family y of sets, the intersection of the family, $\cap y = \{x \mid \forall z(z \in y \rightarrow x \in z)\}$, exists.
IV. (Choice). If x is a collection of disjoint nonempty sets, there is a set y whose intersection with each element of x has exactly one element.
V. (Unordered Pair). For all sets x, y, the set $\{x, y\}$ exists.

Modern set theory has more axioms than these, the additional ones being added for various reasons. Zermelo began the process in 1908 in an accompanying paper offering an axiomatisation of set theory. His new axiomatisation swapped out the redundant intersection axiom for a non-redundant union axiom, added a redundant axiom asserting that for any set x the singleton $\{x\}$ existed, and added the extensionality and infinity axioms. All of these new axioms are ostensibly true about the universe of sets, although one can, as Zermelo would do on occasion, drop the axiom of infinity in order to consider "general set theory", including the hereditarily finite sets as an example. The axiom of infinity, essentially Richard Dedekind's definition of the set of natural numbers, can be viewed either as a true axiom about a universe of sets in which the set of natural numbers exists, or as an axiom necessary for the set theoretic treatment of higher mathematics.

The best example of a set-theoretic axiom added out of necessity is the axiom of replacement added independently by Abraham Adolf Fraenkel and the oft-neglected Thoralf Skolem. The prevailing modern view of set theory is of a cumulative hierarchy growing in stages indexed by ordinal numbers. At stage 0 one has the empty set, at stage $\alpha + 1$ one has the power set of the collection of sets existing at stage α, and, for limit ordinals β, one has the union of all sets existing at stages $\alpha < \beta$. Thus

[4] Gregory H. Moore, *Zermelo's Axiom of Choice: Its Origins, Development, and Influence*, Springer-Verlag, New York, 1982.

$$V_0 = \{\},$$
$$V_{\alpha+1} = P(V_\alpha),$$
$$V_\beta = \bigcup_{\alpha<\beta} V_\alpha, \ \beta \text{ a limit.}$$

Zermelo's initial axiomatisation as well as his general set theory are true in V_ω, whence these theories can only prove the existence of V_n for $n \in \omega$. The axiom of infinity is true in $V_{\omega+\omega}$, which thus models all of Zermelo's axioms, whence, $V_{\omega+\omega}$ not existing in $V_{\omega+\omega}$, cannot be proven to exist in any of Zermelo's systems. This requires a new axiom, the axiom of replacement by which the image of any set under any definite function exists, in this case: $V_{\omega+\omega} = F(\omega)$ where $F(0) = V_\omega$, $F(n+1) = P(F(n))$, and $F(\omega) = \cup_{n\in\omega} F(n)$. And, again, from here one needs replacement to prove the successive existences of $V_{\omega+\omega+\omega}, V_{\omega+\omega+\omega+\omega}, \ldots$

CONVENIENCE. Although nowadays there are nonstandard models of analysis in which infinitesimals and infinite integers exist, this wasn't always the case. Nonetheless, mathematicians freely used infinitesimals and infinite integers as matters of convenience. Gottfried Wilhelm Leibniz even argued that the use of infinitesimals was only a convenience, not actually necessary, and compared them to points at infinity in geometry used to make a more rounded system. Likewise, there is no square root of -1 among the reals, yet mathematicians embraced imaginary numbers willingly.[5] Today we accept the existence of complex numbers and the truths of their properties, but this wasn't always the case and mathematicians nevertheless used such "fictitious" entities freely. Less well-known, but entirely analogous, was a brief period of the non-belief by certain mathematicians in the existence of negative numbers.

In model theory, there is a construction—of *saturated models*—that relies on the generalised continuum hypothesis (GCH). A number of basic results of the theory were easily proven using saturated models and GCH would thus make a nice additional axiom for set theory. Many of these results, however, are essentially arithmetical in nature and Gödel's consistency proof for GCH shows that GCH, along with the axiom of choice, is conservative over the remaining axioms of set theory with respect to arithmetic: any arithmetical statement provable with these new axioms is already provable without them. Thus, GCH and the axiom of choice can be regarded as mere conveniences in proving arithmetic statements.[6]

SYSTEM REFINEMENT. There are several refinements that can be made to an axiomatisation. One can substitute one form of a given axiom for another for the purpose of simplifying derivations. One can sharpen the formulation of an axiom for one reason or another, as when Skolem replaced the "definite property" of Zermelo's separation axiom and of Fraenkel's and Skolem's replacement axiom by the notion

[5] Consider the adage: "the shortest path between two theorems on the real number line passes through the complex plane".

[6] Direct GCH-free proofs of the model-theoretic theorems alluded to have been given, but often require more detailed work than the proofs assuming GCH. So GCH would be a *very* convenient axiom.

of a first-order definable property. These are rather humdrum examples, so I move on. A deeper sort of system refinement is the addition of "missing" axioms. We saw this in the step from Zermelo's analysis of his proof of the Well-Ordering Theorem to his presentation of the full-blown set theory. More famously it can be found in Moritz Pasch's extension of euclidean geometry through the addition of the concept of betweenness to the list of primitive relations of the theory. These axioms are easily justified on the basis of their truth and thus are not what I intend by the phrase "system refinement".

Some "missing" axioms, however, are not merely overlooked truths, but are statements left undecided by the system at hand and are added for the sake of deciding the question.

If by "set" one means "set in the cumulative hierarchy beginning at V_0", then it seems obvious that sets are well-founded and one can add an axiom to this effect as a missing true axiom. But, if one doesn't accept a definite beginning to the hierarchy, one can conceive of non-well-founded sets, e.g., chains $\ldots \in x_{-2} \in x_{-1} \in x_0$ and the axiom of foundation asserting all sets to be well-founded is a false axiom. What justification is there for such an axiom? The answer is simple: the sets of the cumulative hierarchy are vast and all of traditional mathematics can be embedded in them. There is no need to consider other types of sets and the axiom is true of the restricted universe of well-founded sets. In this case, the axiom of foundation is not added as a missing true axiom but as an acknowledgement that one is uninterested in non-well-founded sets.

A second example is afforded by large cardinal axioms. Putting Gödel's rationale for the study of large cardinals aside, there is the simple fact that the Zermelo-Fraenkel-Skolem axioms do not decide whether or not certain large cardinals exist. One can emulate the choice made with the foundation axioms and simply declare there to be no large cardinals at all. Period. Or, one can succumb to the lure of large cardinals and their promise of interesting new constructions and possible new theorems and add these axioms. Opting for their existence, however, involves more than merely making a choice to decide a question; it involves a greater ontological commitment and its concomitant risk. And it does require one to come up with some questionable notion of largeness of a cardinal that would require a new axiom for it to be satisfied. We discuss that shortly.

My third example is, again, the generalised continuum hypothesis. It is true in one universe of sets, namely the *constructible* sets of Gödel's consistency proof, and like the axiom of foundation, the GCH can be added as an axiom true of a large enough universe of sets in which to interpret all of traditional mathematics. My favourite reason for adding GCH to the pantheon of set theoretic axioms, however, is the argument by necessity. Just about any assertion about infinite cardinal arithmetic involving order, addition, or multiplication, is equivalent to the axiom of choice, which makes this much infinite cardinal arithmetic as simple as possible. Exponentiation, however, is not settled by the axiom of choice, but the GCH settles all questions and is the simplest and only known axiom to do so. Thus, I appeal to the authority of Russell's earlier quotation to justify GCH as such a necessary axiom for set theory.

ANALOGY. Large cardinals are also proposed by analogy, the axiom that some cardinal number behaves in some ways like ω. One sometimes represents this as completing a proportion: κ is a large cardinal if it satisfies

$$1 : \omega :: \omega : \kappa.$$

The ratios exhibited are not really meaningful, but consider, viewed as ordinal numbers, the cardinals $1 = \{0\}$ and $\omega = \{0, 1, 2, \ldots\}$ are both closed under addition and multiplication. Any κ also satisfying this is an infinite cardinal. 1 and ω are also closed under exponentiation and replacement. Larger cardinals κ with this property are inaccessible.

Without the argument of convenience, one could also justify the use of infinitesimals as being real-like numbers satisfying properties analogous to those of the real numbers. These two examples are thus of the form: suppose new entities satisfying known properties exist; let us study them. In this I suppose the new axioms could be considered formal developments that will either turn out to be inconsistent and dropped or to be successful and accepted as axioms about genuine mathematical objects. But one can imagine other types of axioms by analogy. For example, and here I do not know the history, one can view the theory of modules over various rings as analogically derived from the theory of vector spaces over fields.

PURE FORMALISM. My final category could be labelled "Other" or "Arbitrary". I have however chosen to label it "Pure Formalism" to acknowledge that there is some rationale behind the choices of such axioms. These axiom systems, while not attempting to describe some existing structure or structures, are attempts to describe something and are not merely "arbitrary". I can think of four examples: the various noneuclidean geometries of the 18th and 19th centuries, the lambda calculus of Alonzo Church, the set theories of Willard van Orman Quine, and the linear logic of Jean-Yves Girard.

Today we have models of noneuclidean geometry, so we know the axioms are true about something. Initially, however, the theories were developed formally, an axiom chosen to contradict the parallel postulate and its consequences studied in the hope of finding a contradiction. In this way, in the 18th century Gerolamo Saccheri found sufficiently counter-intuitive consequences to convince himself he had found such a contradiction. Realising an outright contradiction had not been found, others continued the development until the likes of Gauss, Janos Bolyai, and Nikolai Ivanovich Lobachevsky recognised noneuclidean geometry to be a fully consistent theory, but one not known to be true of anything. It was a purely formal exercise in axiomatics until the discovery of actual models by the likes of Felix Klein.

Church's lambda calculus was likewise formal, not a theory about an existing structure, but a calculus formed to analyse substitution in the presence of bound variables, with logical operators thrown in willy-nilly. The system was quickly shown to be inconsistent and thus it was a theory about nothing. Church was able to rescue his system by dropping the propositional calculus and developing the so-reduced lambda calculus as a model of computation. As such it was still a formal exercise, but a quite successful one.

Quine's misleadingly named systems New Foundations (NF) and Mathematical Logic (ML) of set theory were in no way attempts to axiomatise set theory by listing evident properties of sets, but were, rather, attempts to develop a set theory with a universal set of all sets by restricting the formulæ to which unbounded comprehension applied. That the enterprise was purely formal became undeniably evident when the original system was shown inconsistent, whence not any form of foundation, new or otherwise. As far as I am aware, the current incarnations of NF and ML are not known to be consistent and are thus still purely formal in the much maligned sense of the "formula game" wrongly associated with the name of David Hilbert.

And Girard's system, as near as I can make out, was motivated as follows. There are two major variants of sequent calculi. Combining these into a single system with two copies of each propositional connective yields a general system with no immediate semantics, but, like Church's lambda calculus, possibly great applicability. Not having followed the development in the decades since the birth of linear logic, I have no idea of the current state of affairs regarding it and can only cite it as a known example of a formally based theory.

So we see that mathematical practice does not follow the Greek paradigm of always choosing one's axioms because of their evident truth; axioms may be chosen for a variety of reasons which may or may not overlap in particular cases. And their justifications can even change over time. As repugnant to one's higher susceptibilities as it may be, the most all-encompassing description of the reason for choosing axioms may well be Hilbert's phrase "arbitrarily given". I am now inclined to believe that one genuinely does have complete freedom in choosing a set of axioms to study, but I also believe that one should be aware that the significance of the set chosen is guaranteed by what it does, most obviously measured by its connexions with the rest of mathematics.

Acknowledgements The present note is a concise version of an identically titled section from the chapter on the axiomatic method in my book *Adventures in Formalism* (College Publications, London, 2012). The interested reader is referred to this book for much elaboration.

Craig Smoryński earned his Ph.D. in 1973 with a thesis on Kripke models for intuitionistic systems—propositional (intermediate logics), first order (decidability and undecidability results), and arithmetic. This background proved useful later in Amsterdam when he joined Dick de Jongh in launching provability logic, publishing what remains the only research monograph in the field (*Self-Reference and Modal Logic*), despite tremendous strides made by Dutch, Italian, and Russian logicians. In recent years he has turned his attention to matters more historical (*Chapters in Probability, A Treatise on the Binomial Theorem, and MVT: A Most Valuable Theorem*) and general (*Adventures in Formalism* and *Mathematical Problems: An Essay on Their Nature* and *Importance*).

Chapter 11
Panel Discussion on the Foundations of Mathematics

Fernando Ferreira

Abstract On the afternoon of October 11, 2017, in the grand hall of Academia das Ciências de Lisboa, it took place a panel discussion on the Foundations of Mathematics. The discussion was part of the second Axiomatic Thinking Meeting, the one in Lisbon. The invited panelists were Peter Koellner, Michael Rathjen and Mark van Atten, and the moderator was Fernando Ferreira.

F. Ferreira (✉)
Universidade de Lisboa, Lisboa, Portugal
e-mail: fjferreira@fc.ul.pt

F. Ferreira et al. (eds.), *Axiomatic Thinking I*,
https://doi.org/10.1007/978-3-030-77657-2_11

11.1 Introduction

On the afternoon of October 11, 2017, in the grand hall of Academia das Ciências de Lisboa, it took place a panel discussion on the Foundations of Mathematics. The discussion was part of the second Axiomatic Thinking Meeting, the one in Lisbon. The invited panelists were Peter Koellner, Michael Rathjen and Mark van Atten, and the moderator was Fernando Ferreira. The panel discussion was recorded in a mobile phone and transcribed by António Fernandes (whom we very much thank for this work). Some parts of the audio file were barely, or none at all, audible. So, there are some gaps and some minor guesswork in the transcription. However, with the exception of Michael Detlefsen, all the known intervenients saw the transcription and made some adjustments. We asked the intervenients to keep these adjustments to a minimum, in order not to spoil the oral, spontaneous character of the interventions, asking them not to succumb to the temptation to polish the interventions as if they were miniature essays. By and large, this was accomplished.

Professor Detlefsen did not correct his part of the transcription because, unfortunately, he passed away meanwhile. We were unsure whether to insert his interventions under his name, and so we consulted his colleague Professor Patricia Blanchette. She encouraged us to publish his contributions to the discussion and gave some small editing suggestions, primarily about punctuation. This panel discussion is dedicated to Professor Detlefsen.

11.2 Discussion

Fernando Ferreira. First of all, I would like to thank the panelists for agreeing to be part of this discussion about the current foundations of mathematics. I will give five minutes to each panelist, maybe in the order of the talks: first Peter Koellner, then Michael Rathjen and then Mark van Atten. And then I will open the floor for questions and see where it leads. So, maybe, Peter you can start.

Peter Koellner. In terms of the foundations of mathematics, I think we should let a thousand flowers bloom, and I'm interested in many different aspects—ordinal analysis, reverse mathematics, getting more information through constructive proofs, ...—but, here, I will focus on an area that is of particular interest to me—the one that I spoke on earlier. I will try to focus on matters connected with Hilbert. I'll start from the concrete and move out to the abstract.

At the concrete level, I'm interested in consistency questions such as "How far does the interpretability hierarchy go?" I want to know which systems are consistent, and in saying that I do not mean that I want an oracle for consistency—that would be too much to ask for, that would involve an oracle for Π^0_1-truth. But I want to know whether some special markers in the hierarchy that are high up are consistent, like ZF + Berkeley Cardinals. And, related to that, I want to know whether the other future holds, where the Ultimate-L Conjecture holds, and that is another concrete statement because it is just a conjecture about what is provable in a formal system.

Moving on to the less concrete realm, suppose that it turns out that the good future pans out and the Ultimate-L Conjecture holds, and so this top level of the consistency hierarchy—where we have large cardinals that violate choice – evaporates. Then we are in a really good situation with regard to resolving incompleteness. There are two kinds of incompleteness, there is the vertical independence of the incompleteness theorems and there is the horizontal independence that arises in forcing, e.g. the independence of CH. They are both problems, but the second is much more of a problem. We have a harder time dealing with CH than we do with the consistency sentences that arise in the incompleteness theorems. So, for example, PA can't prove Con(PA) but we have a better grip on Con(PA) than we do on CH. So, the virtue of this first future panning out is that you have an axiom "scheme" (where "scheme" is in quotes here—it is not like mathematical induction) ZFC + V = Ultimate-L + "Large cardinals". The second axiom—V = Ultimate-L—would wipe out horizontal independence and would leave us with just vertical independence, and to deal with that we could just keep climbing up the large cardinal hierarchy. So we would have a good grip on the answer to these questions, we would have a theory which was effectively complete.

But then going even more abstract, once we get a theory that is doing a good job handling independent statements, once we get a theory which is effectively complete, there is the question of whether it is true and I would like then, at that stage, to know that and that would involve mining the consequences of the theory.

Fernando Ferreira. Thank you very much. So, maybe, now you Michael.

Michael Rathjen. Definitely I agree with Peter's description: let a thousand flowers bloom. This is also a very Hilbertian point of view. I think Hilbert didn't want to restrict mathematics in any way. Mathematics has fared very well, so that there was never a really a good reason for restricting mathematics by saying that we shouldn't go there. Of course, you might hit upon contradictions and then would have to backtrack. Hilbert's ideal elements can be viewed as promoting openness. They are very important to mathematics. This is the idea that we use ideal elements to give structure to your mathematical experience and they can actually guide us to success. In some sense the ideal elements provide a kind of a story that humans are craving for in order to make sense of very complicated data and to find structure in them. They play a really important role, and then of course afterwards, there one almost always had this amazing result that when it came down to very concrete mathematical results in number theory—not the ones that came from Gödel's encoding of consistency—that these results could be proved without the ideal elements. I think it is a very amazing story which has not really been elucidated. I mean, we have this Gödel hierarchy and when we use it, very often we find out afterwards that we no longer need it. That's fine, but I think there is a need for an explanation of this phenomenon. When we look at the reverse mathematics program, it turns out that most of the time mathematics can be done in the early systems of reverse mathematics, i.e., mostly in RCA_0 and WKL_0 and stretching to ACA_0, i.e., up to the level of Peano arithmetic, but not much beyond that. I should say that there are exceptions to this, though. In the story of the big five systems, one exception is definitely the Graph Minor theorem. The Graph Minor theorem is an amazing theorem proved by Robertson-Seymour in more than 20 papers and running up to 600 pages, and it is probably also one of the last great ivory tower enterprises, where some people talk to each other sitting in an ivory tower and come up with a proof of a great theorem. Another example is Wiles' proof of the Taniyama-Shimura conjecture. This might change in the future. I think it will change in the future, perhaps a lot. We can only guess what future changes might be by extrapolating in a linear way, but there might be really exponential changes in the way we do mathematics. One of the great things about the Graph Minor Theorem is that it also has a lot of predictive power. It predicts that if you have a graph problem that is closed under taking minors then there is an algorithm associated with it. So, for instance, the knotless graphs are closed under taking minors. The Graph Minor theorem predicts that there are algorithms—that run in cubic time—which can decide the property of knotlessness, but currently we don't know any such algorithm, as far as I know. This is an example of how powerful the use of ideal elements can be in that they predict the existence of some concrete things, so we can go about searching for them. If we didn't know that they existed, we might not be able to find them because we wouldn't have the confidence and stamina to search for them. So, this was one remark. The other relates to what Peter said about his interests in the hierarchy of very large cardinals. I think that the mystery of largeness and strong set notions comes up earlier in areas which have been very much jumped over. For instance, in ordinal analysis people have been trying to do ordinal analyses of strong systems but basically the current state is frozen somewhere around Π^1_2-comprehension. The consistency of analysis, of second-order arithmetic as a formal system, was one of Hilbert's problems. It was a kind of holy grail problem, and for some people still is. It has not been pursued beyond Π^1_2-comprehension, and I think this is an area which should also be explored. There is a long stretch between Π^1_2-comprehension and large cardinals. This kind of neverland, unexplored land, is starting a bit beyond the level of Π^1_2-comprehension. In ordinal analysis one attempts to assign meaning to these impredicative axioms, positively constructing a mini-universe by means of which one can somehow understand difficult impredicative comprehension at the levels of Π^1_3 and Π^1_4-comprehension and so on. Currently nobody has an idea. There must be a generic case, though, that is, one probably doesn't always have to come up with a new idea, e.g. when one goes from Π^1_7 to Π^1_8. There must be a jump at some point that allows one to understand the full hierarchy. My conjecture is that the generic case is the Π^1_3-comprehension case. This is something that should be looked at. And then, for proof-theorists, there are interesting things related to univalent foundations with a lot of potential. Questions like the identity of proofs bringing together proof theory and algebraic topology, and questions about the compatibility of univalence with resizing rules. There is a lot of stuff going on. And then just one remark about the future of mathematics. For instance, DARPA funded research projects are targeting the interface between human brains and computers to enhance the human brain. This might also play a substantial role in the future of mathematics and could conceivably fundamentally change the way how we go about doing mathematics.

Unknown. Is it also part of that project to enhance computers? By connecting them to humans?

Michael Rathjen. Well, this will happen anyway. For eight years I was on the scientific board of Oberwolfach, where you get to read a lot of research proposals. I got the impression that research hypotheses in some areas are very much informed by the use of computers. Some younger people use computer tools quite a lot, which made me feel a bit like a dinosaur.

Fernando Ferreira. Thank you. Now Mark.

Mark van Atten. (...) A concrete wish that I have, at least for philosophy of mathematics, is that Kreisel's papers will be collected and made accessible, preferably on the internet. This should be realized as soon as possible. Speaking more generally, looking at the history of philosophy of mathematics, I think there is a tendency to say nowadays that the foundational debate between, let us say, the classical positions: formalism, intuitionism, logicism, platonism, is really behind us and we should turn to mathematical practice, mathematical experience, history of mathematics. I think that is a mistake. Of course, historically speaking the debate has calmed down but I think this is because of psychological and sociological reasons, but certainly not for philosophical reasons. Because I think that the philosophical questions that the participants at the time were concerned with are very much philosophical questions now. We should not forget that these philosophical questions that motivated the original debate are still open. And one important reason for these standing questions is that if you say we should turn away from these foundational issues and study more of mathematical experience, one fundamental question for debate is: what counts as mathematical experience? What makes certain experiences mathematical? I think that much more work can be done on that.

Fernando Ferreira. Thank you. Now I give the floor to the audience. I would just ask people to identify themselves. If you want to make a question, please say your name first.

José Ferreirós. One question is that if one compares the situation today with the situation of one hundred years ago, and even more so if you go backwards to the nineteenth century, it seems to me, almost everybody that comes to my mind, Gauss, Weierstrass, even Hilbert, I think, had this picture that one finds one uniform foundation for mathematics, a single basic system. Nowadays it is different, there is a pluralistic thinking. Why do you think that this happened?

Michael Rathjen. Could you give an example, where pluralistic thinking has been employed?

José Ferreirós. I mean, all of you have this idea of letting a thousand flowers bloom. (...) The way I understand Hilbert, in this paper Axiomatic Thinking, I don't think they were aiming at diverse foundational systems, rather they had this belief that you will find a single system. Is it because of Gödel that we have changed our minds?

Peter Koellner. I think it is important to distinguish between two senses of the phrase "let a thousand flowers bloom". I want to explain in what sense I used that phrase. I wasn't using it in the post-modern "truth is relative" sense, where you can have this framework, you can have that framework and truth is local to the framework and there is no outside objective matter when it comes to truth. (Like Π^0_1-sentences. I will not go relativistic about Π^0_1-sentences.) And the statement that I made about "let a thousand flowers bloom" was really just that there are a lot of interesting aspects of mathematical logic which, if you don't take them as complete on their own, can coexist together. So, when I said that there is a lot of information that we can get from constructive proofs, then, in saying that I'm letting two flowers bloom: constructive proofs and nonconstructive ones, and so I'm viewing constructive proofs as part of nonconstructive mathematics. So, I guess, in saying "let a thousand flowers bloom" one region of the garden that I'm trying to keep in existence is higher set theory. Certainly, it wasn't a truth-relativist statement.

Gregor Schiemann. A close reading of Hilbert's lecture shows that there are not only different versions of unity in it. He acknowledges also an astonishing amount of plurality in the sciences. He speaks of science only in the plural. I just want to remark: It is not clear how far Hilbert goes with his claim of unity.

Wolfram Pohlers. At the time of Hilbert, Kronecker and so on, the foundational problems of mathematics were very vivid. To my view, it is no longer vivid outside the community of mathematical logic. For everyday mathematicians (if I may call them so) these are problems that do not play any role any longer. What is the reason for that? Are there any answers to that?

Michael Rathjen. As to plurality, I certainly did not mean this in the sense of French postmodernism or Feyerabend's "anything goes".

I think of plurality in the sense that one should have a science base where there is a certain amount of plurality. If everyone is following the same track, we will miss out on important aspects of reality and this is not healthy for science. The thing is that some dominating people or groups can create a certain narrative and the young ones will jump on the bandwagon. But I think that the truth is rarely pure and almost never simple. Sometimes we look at this incompleteness result, due to Gödel, which is very important, and we say, OK, this refuted Hilbert, but one should also not forget that there are other areas where there is a lot of completeness. The history of mathematics, for instance, is also a history of incompleteness where one starts with a certain realm of mathematical objects like numbers, but then it turns out that they do not suffice in that certain equations one wants to solve have no solution. And this has been going on and is totally different from Gödel's incompleteness. For instance, in model theory, some people say that Gödel does not really matter because there are lots of complete theories. Sometimes people gleefully say, oh yes, Hilbert thought that a first-order system could be complete. There are lots of complete first-order systems—Euclidean geometry, for instance. Anyhow, such complete theories play an important role in model theory. So, I think that plurality should mean to have an open science base, and not to have any group dominating totally. I think this would be also in harmony with what Hilbert thought. I don't think that Hilbert was after a once and for all system. The main thing is that we know how to move between systems. Though we have a plurality of different frameworks, we can reasonably and rationally move between them. This is totally different from postmodernism.

Gerhard Heinzmann. Hilbert's program is a surely promising justification for strictly formalized mathematics. But do you think it gives reasonable insights for understanding mathematics? I mean this is a better question, already raised by Poincaré, than saying: oh yes, we should return to old problems and results not from the point of view of comprehension but of justification. The question of understanding is, of course, the question of the practical turn in the philosophy of mathematics, i.e. not simply to ask for a proof, but for understanding better, i.e. for a good proof. In other words, can Hilbert's program contribute not only to the justification but also to the explanation of mathematics?

Mark van Atten. Yes, the differences between these kinds of proofs that you mentioned, they really matter. There is the basic question underlying all of them: What is really a proof? There are proofs that you really do not understand but you recognise to be a proof. So that question, foundationally, comes first. The individual philosopher of mathematics should of course decide for him- or herself which question to work on. I think that formalization is very fruitful for organizing and presenting knowledge. Also for understanding. When it is possible to formalize a body of knowledge ...

Gerhard Heinzmann. It is not enough.

Mark van Atten. ... it may make things easier to understand, for one has a sharper description of at least certain aspects of the objects and properties involved.

Peter Koellner. I have a comment on that. A lot of the original foundational discussions involved classical questions of epistemology and metaphysics. People wanted to know how we could know mathematical propositions especially in light of the fact that they deal with abstract objects that we do not have causal contact with. If Logicism had turned out to be true, then that would have been great because we could then "define away" all the objects—they would be virtual objects—and you would solve the access problem, and you would have a pretty good answer to the epistemological problem because then the principles of mathematics would reduce to analytic truths. And if Hilbert's program had worked out that would also have had great payoffs in epistemology and metaphysics. Things did not turn out that way. We still have the epistemological problems and the metaphysical problems. It is just that they are now a lot harder. Instead of having one framework like Logicism where it all reduces to analyticity, we have got a hierarchy of frameworks and the epistemological problems become harder and harder, and the same is true for the metaphysical problems as you go further and further up. To me, that makes it more interesting. Epistemology and metaphysics are just really hard subjects. If you go to epistemologists and you say: have you solved your first problem?—the problem of induction—they will say: no. I think it is a very difficult subject. The same thing holds in physics. After Newton people got really optimistic; they started talking about "attraction of ideas", based on the model of gravitational attraction, and they started to think that pretty soon you would be mechanizing everything. La Mettrie wrote about the Machine Man. A lot of people were overly optimistic. It did not turn out that way, and got much more interesting – relativity, quantum mechanics, ... and that raises with it very difficult epistemological and metaphysical problems. So, I think there is a parallel.

Reinhard Kahle. I will propose two other questions to the panel. I would say that it is clear why mathematicians are not very interested in foundational problems.

We have set theory and the foundational problems are outsourced to set theory. If somebody comes and says to a mathematician: "well this is good but we don't know whether it is consistent; so, is it meaningful?", the mathematician answers: "I know that I can formalize it in set theory; so, about consistency, speak to the set theorists." Now, set theory actually provides a universal language for all of mathematics. This is not an ontological claim about mathematics; when mathematicians work in algebra or analysis, they do not work in any specific set theory; but we can formalize the work in set theory, and so we can reduce the consistency questions to set theory. I think that nobody questions the ability of set theory to be the universal language of mathematics, even if we are, by no means, obliged to put everything into set theory.

Here my two questions:

Is there any chance, that type theory could equally be a universal language for all parts of mathematics? It is clear that univalent foundations are very successful for certain specific problems, but has it the potential for being as universal as set theory? And for Mark: Would the intuitionistic approach, intuitionistic mathematics, be as universal as set theory?

We have set theory with these very, very large cardinals. Do they have really relevance for the working mathematicians, or is it just a game on consistency, in some spheres which never could impress a mathematician?

Michael Rathjen. As far as type theory goes, this is quite a difficult language to work with. Part of the community is also in Computer Science with different standards. Since I was not brought up in this community, I sometimes have problems with their papers. I have come across some, where they claim to be able to prove a certain result, e.g. that something can be done in a certain type system, and, as they go about the proof, they change the system. I also recall many discussions as to whether specific rules are allowed, e.g. the ξ-rule. So, you often don't know where you are. What I want to say is that type systems can be very difficult to work with as there is a lack of robustness and the community has not settled on a fixed system. This is totally different from set theory where everything is very simple and robust. Of course, there are systems like Coq, they get used and apparently are very successful. As to the question whether type theory could be a background system for mathematicians and whether they want to learn it, I think the current hype is probably overrated, so I have my doubts. On the other hand, who knows. As they say, it is very hard to make predictions, especially about the future. I think that set theory has an extremely simple language and is still very successful. But success could also lead in another direction. For instance, there was this discussion about the axiom of choice in 1905 involving the famous French analysts, and the verdict was basically that the axiom of choice that Zermelo had introduced is something that does not belong in mathematics. But then in 1930 a paper by Teichmüller describes the role of the axiom of choice in algebra, showing that it is just very useful, unifying a lot of things. Success is something that is very important and can drive the enterprise in any direction. Mathematicians are not so much concerned about foundations when their area thrives; if it is successful it is not that useful to raise foundational questions if you just want to do your mathematics.

But, on the other hand, there are still the philosophical questions that Peter and Mark have spoken about. Maybe there's a question for the audience as we have several experts on the philosophy of mathematics in the audience. My question is: Have we really made significant progress in the philosophy of mathematics since the days of the renaissance guys, like Hilbert, Brouwer and Russell. Of course, we know a lot more, and have very subtle distinctions these days. Today, I saw on one of the slides a reference to a recent book by Ian Hacking. In it he seems to adopt a position of humility vis-à-vis the great philosophies of mathematics of the past—often addressed as logicism, formalism and intuitionism—in effect saying: who am I to judge? It would be interesting to hear what you have to say about progress in the philosophy of mathematics.

Now, going back to Hilbert, he is also known for his no *ignorabimus* stance. *Ignorabimus* means that there are limits to what humans can understand, a position that Emil du Bois-Reymond put forward in the 19th century in Germany. One of the things that he thought humans could never understand was consciousness. In recent books by Chomsky, he distinguishes between problems and mysteries, where the latter remain ungraspable to human minds, and he also lists consciousness as a mystery. It is perhaps more exciting to align with a Hilbertian stance, remaining optimistic as to the solvability of mysteries. Anyway, I'd be interested to hear your thoughts on this.

Mark van Atten. Of course you have intuitionistic views on set theory. It depends. Do you look at set theory as a formal theory or as contentual theory? As a contentual theory, classical set theory is not true. It talks about things that do not exist because they cannot be constructed. But, of course, as soon as you speak of it as a formal theory, you can say: fine. Most classical mathematicians, at least those foundationally minded, are somehow formalists ... and from the intuitionistic point of view that is fine. As a historical note, in the 1920's Brouwer, in print, expressed optimism about Hilbert's program twice. But he also realized that he and Hilbert would disagree about the value of that expected success. There is another reason, which Hermann Weyl remarked already at the time. It seems that classical set theory leads you to see everything as discrete, and Weyl said that if you do that, then you will never have a satisfactory theory of the intuitive continuum. This is certainly also Brouwer's take. Then again, Hardy, for example, said: "Then so be it, I prefer my discrete theory". Hardy acknowledged the point but, for pragmatic reasons perhaps, he would take the discrete theory.

Peter Koellner. I will say something about these two questions, the first question being about the supposed decline in interest of mathematicians in foundational issues and the second question being about the relevance of higher set theory to what core mathematicians do. These are both sociological questions, so I will be acting as a sociologist now even though I am not an expert, so these are just my impressions. I think there has been an increasing compartmentalization of the disciplines. The people we were talking about—they were really renaissance people: Russell, Frege, Hilbert ... Hilbert was involved in everything. There has been an increasing compartmentalization of the disciplines, largely due to the fact that with the advancement of the sciences each one gets more difficult and human time is finite (especially with administration and things like that), so it is harder and harder to keep abreast of other fields; and one of these fields happens to be foundations of mathematics. So, if you are a working mathematician when will you be interested in the foundations of mathematics? Well, one area in which you will be interested in is this: if you are working on a problem and the problem is not resolvable on the basis of the axioms you are using, then you want to know. And that is when mathematicians generally have their interest piqued. Shelah proved that the Whitehead problem for groups is independent of ZFC, and mathematicians were interested because people were working on it. The same thing happened with Kaplansky's conjecture. The Calkin algebra is a recent example. Connecting this up with the relevance of higher set theory—the set theorist just showed: "Look, you can beat your head against the problem forever, but you are not using the principles that would resolve it". And typically what they do is say: "it is not a mathematical problem anymore". That is what they said about the Whitehead problem. It is not math anymore. I think that is a disingenuous response which is connected with this compartmentalization. I do not think that people back in Hilbert's time would have said that it is not a mathematical problem anymore.

Michael Rathjen. It is interesting that you bring up the Calkin algebra. In the case of the Whitehead problem there was some resistance. But in the case of the Calkin algebra, this has actually galvanized the C*-community. They like this stuff. They also seem to like it in a certain direction you might not like as they tend to think that a problem is solved if you can solve it with the assumption $V = L$. These are some observations that I learned from Nik Weaver's recent book on forcing and the time on the Oberwolfach committee. This just goes to show that things can change. It's never just one direction, over time interests can shift. Also, in Proof Theory, proof mining has spread into areas of analysis, approximation theory and more.

Peter Koellner. If I can build on that, I was only referring to their interest in foundational issues. Then there is the question about their interest in mathematical logic. Mathematicians are pretty interested in parts of mathematical logic. They are interested in model theory, in parts of Farah's work in set theory that have had application in operator algebras (he recently gave a talk at the ICM, which shows a certain amount of interest). And there is a massive amount of work on top of these independence results where they are actually applying set theoretical techniques and solving problems that they were not able to solve before. There is a flow of ideas from set theory to other areas of mathematics and that is good and healthy. I am a little concerned when people start to say that there must exist a flow. I think that a part of letting a thousand flowers bloom is persisting through periods where there is not a lot of flow. Is there any flow from number theory to set theory? No. I will not criticise number theory for not having flow in that direction. And likewise in physics. Does the work on black hole thermodynamics have any relevance to someone working on fluid mechanics? No. Flow is great when it naturally arises. I am just pointing out that there is some flow from areas of mathematical logic to more mainstream mathematics.

Unknown. (...) It seems to me that there is a great difference between the formalizing of sciences with mathematical proofs, and what we find in them with regard to their content. (...) So, this is one, so to say, perspective. The other perspective is that we formalize science and mathematics, we do not have this content.

Michael Detlefsen. (...) I want to comment on the *ignorabimus* remark. It seems to me that it is one of the shadiest views ever in the philosophy of mathematics. (...) Because Hilbert has classified the solution to a problem. I'd like to have the solution to a certain class of polynomials; I know what to do; I will create a number that multiplied by itself is negative. What the hell kind of solution is that? What does it really show? It is sort of a determination (...) It counts as a solution anything we can get, by any means. To me, it is not a very robust notion of solution. We say that in mathematics there is no *ignorabimus* because we will count as a solution anything we can get by. I am being sarcastic, but I think there is a serious point behind this. Also there is this little piece of textual material that needs to be brought into this particular discussion, and that it counts as solving a problem, showing its unsolvability. (...) I quote Hilbert here: "Every definite mathematical problem must necessarily be susceptible of an exact settlement, either in the form of an actual answer to the question, or by a proof of the impossibility of its solution and therewith the necessary failure of all attempts." If you do that, you solve the problem. (...) I am more interested in what seems to be a pretty shady history of counting things as solutions that I think most (...).

Peter Koellner. Well I was using it in a specific sense, not in the sense of cooking up something to satisfy the problem, I was using it in the sense where you already have a good understanding of the domain of the objects you are talking about, you already have a good understanding of the propositions concerning that domain, like number theory, and the belief that if you ask a question—for instance about number theory—we will be able to answer it. The kind of thing that Gödel held. Just for reasons of limitation, we cannot say that we will be able to answer all questions of number theory because we are extremely finite, most of those propositions have 10^{10} symbols and we cannot even understand them. So we are only concerned with things that we—with our limited, feasible resources—could ask. And with the idea that if we can ask it and just work hard enough we will figure it out. A kind of rational faith.

Michael Detlefsen. If you want to understand what Hilbert said, you have to be serious about consistency being enough. So, if you consider the axioms governing our thinking about complex numbers, they are consistent. Even if I get the smoggiest things. How could it be that I get negative squares? Let me say, because these axioms are consistent. And it allows solving equations that you cannot solve without complex numbers. That's pretty much it.

Peter Koellner. I am not an Hilbert scholar but I would like to hear what other people think: was Hilbert a relativist about arithmetical sentences?

Michael Detlefsen. What do you mean by that?

Peter Koellner. Well, you said that for him it is good enough to have the consistency of the system. So take PA + ϕ—where ϕ is some arithmetical Orey sentence—and PA + $\neg\phi$. They are both consistent. So would he say in that situation that we are in a third category were you have answered the question by giving a meta-answer, namely, that there is no answer?

Michael Detlefsen. I do not know the answer to that question. But there seems to be a difference by virtue of that he certainly believed that the addition of numbers, negative squares, he thought that was one of the crowning achievements of the idea that is the prime example how we mathematicians go about solving a problem.

Wilfried Sieg. I suppose that there is also a need to understand that the construction of (...) models of axiom systems has faced significant problems early on.

Michael Detlefsen. That's a complicated method.

Steve Simpson. That's an interesting example, I mean because these weak systems based on PRA are strong enough to formalise a lot of complex variable theory. There is a distinction between analytic and elementary proofs in number theory but it is pretty safe to say that almost every theorem of number theory whose proof uses complex analysis has an elementary proof in PRA. This makes the complex numbers a kind of ideal elements and they are harmless, and that's the idea of Hilbert's program. It is a good example of what is involved in Hilbert's program, what he was looking for. We introduce the complex numbers but that is harmless because in the end this introduction is conservative.

Michael Detlefsen. (...)

Steve Simpson. If a universal number theoretic statement is consistent with, say, the Peano axioms or PRA or even weaker systems, then it is true.

Peter Koellner. Yes. If the Riemann hypothesis is consistent with PA, then the Riemann hypothesis is true. Because if ϕ is Π^0_1 and is consistent with PA, then $\neg\phi$ can't be true, because $\neg\phi$ is Σ^0_1 and PA is Σ^0_1-complete, and would prove $\neg\phi$, contradicting the consistency of PA+ϕ.

Steve Simpson. Hilbert was right about that when it applies to universal statements. That if a universal number theoretical statement is consistent, then it is true.

José Ferreirós. Concerning Hilbert, I think it's important to consider his writings that are not on the topic of foundations. There is this lecture from 1920, that was published some years ago, where he has a striking statement: "Mathematics is not like a game whose tasks are determined by arbitrarily stipulated rules. Rather, it is a conceptual system possessing internal necessity that can only be so and by no means otherwise." So he insists on the content of mathematics, and says that the development of mathematics is driven by necessity. Very strong statements.

Michael Detlefsen. (...) Don't forget that it is an activity that consists in operating with meaningless symbols and still not a game. That is why he wants to prove consistency. You do not prove the consistency of the rules of chess.

José Ferreirós. He also said that proving consistency cannot solve all the interesting epistemological problems about mathematics (...)

Unknown. I can see that mathematicians are not at the moment interested in forms of complexity, in constructing more complex systems. But today with the new types of computing, we have this induction where the problem is finding the function because we have a set of big data. There are different types, you do not necessarily know the type there is but you are trying to find the regularity of the structure in the chaos of immense data. You do not have *a priori* the functions that you are trying to discover. Do you say that there are still mathematical problems in trying to find this? If we find one, you may have different types of functions. Wouldn't it be a mathematical question to see that the different functions are consistent among themselves. You are not necessarily talking about the content but the mathematical structure of discovery itself and the value of the function discovered. Machine learning processes ... is discovering the functions (...)

Peter Koellner. Is this question stimulated by that earlier statement I made that in general mathematicians don't seem to be as interested in the foundations of mathematics as they used to be?

Same unknown. I don't mean the foundations, but the contents.

Peter Koellner. I did not say that they are not interested in the content. They are interested in the content, I would think.

Wilfried Sieg. Axiomatic thinking is supposed to be the unifying theme of our conference. Axiomatization is for Hilbert the particular method of approaching issues in mathematics itself, in the foundation of physics, indeed in the foundations of any subject that can be rigorously formulated. And, I believe, we have to be very careful when speaking about Hilbert's thinking on the issue in this particular way. There is a quite systematic and dramatic development in his thinking. It does not start in the 19th century and go uniformly to the 20th century; on the contrary, there are big developments in between. For example, the form of a logical calculus that he used in the 1920s, that he learned late, namely, in 1917/18 from Principia Mathematica. So, to think that Hilbert's worries about the axiomatic method and consistency from the late 1800s are similar to those from the 1920s is to miss out on very important developments in mathematical logic. After all, Hilbert's finitist program was invented only in 1922. It is not something from the 1890s. One has to be very careful to formulate issues occurring in the 1920s and then to compare them to whatever was taking place in 1895. One has to take into account the complexity of the development of Hilbert's thinking and that of mathematics and of mathematical logic.

Fernando Ferreira. Thank you for this word of caution about Hilbert. Unless you want to say something else, maybe we can stop here. Let's thank the panelists and the audience. It was an interesting discussion.

Printed in the United States
by Baker & Taylor Publisher Services